INSTITUTIONAL INTEGRITY IN HEALTH CARE

Philosophy and Medicine

VOLUME 79

INSTITUTIONAL INTEGRITY IN HEALTH CARE

Edited by

ANA SMITH ILTIS

Saint Louis University,
St. Louis, Missouri

KLUWER ACADEMIC PUBLISHERS

DORDRECHT / BOSTON / LONDON

A C.I.P. Catalogue record for this book is available from the Library of Congress.

ISBN 1-4020-1782-0

Published by Kluwer Academic Publishers,
P.O. Box 17, 3300 AA Dordrecht, The Netherlands.

Sold and distributed in North, Central and South America
by Kluwer Academic Publishers,
101 Philip Drive, Norwell, MA 02061, U.S.A.

In all other countries, sold and distributed
by Kluwer Academic Publishers, Distribution Center,
P.O. Box 322, 3300 AH Dordrecht, The Netherlands.

Printed on acid-free paper

Printed in the Netherlands.

TABLE OF CONTENTS

ACKNOWLEDGEMENTS

I would like to thank H. Tristram Engelhardt, Jr., Editor of the *Philosophy and Medicine* series, and Lisa M. Rasmussen, Managing Editor of the series, both of whom provided guidance throughout the preparation of this volume. I would also like to thank Jerome Schmelzer for his editorial assistance.

I also wish to express my gratitude to the Center for Health Care Ethics, the Graduate School, and my colleagues at Saint Louis University for their ongoing support.

Finally, I wish to express special thanks to my husband, Steven for his long-standing love and support as well as for the many hours he spent editing and formatting this volume.

Ana Smith Iltis
Saint Louis University

ANA SMITH ILTIS

INSTITUTIONAL INTEGRITY IN HEALTH CARE: ESSENTIAL FOR ORGANIZATIONAL ETHICS

Health care delivery has changed radically. Physicians no longer are independent agents who develop relationships with patients and, together with patients, make decisions about appropriate courses of treatment. Instead, health care organizations or institutions, including hospitals and health insurance companies, are powerful entities who greatly influence health care delivery, including decisions about who receives care, what type of care they receive, and when they receive it. Because they have entered the social world of relationships, health care organizations can now be seen as bearers of moral obligations subject to the possibility of acting in ways that are good and bad, right and wrong. As such, health care organizations are, in the words of Peter French, "proper subjects ... of moral assessment" (2003, p. 20). The need to focus on the moral obligations of health care organizations has not gone unnoticed. Ezekiel Emmanuel observed almost a decade ago that

> Managed care has fundamentally changed the nature of medical ethical issues. They no longer arise in the context of individual patients and physicians. Instead they arise in the context of complex institutions that establish an organizational framework in which these ethical issues arise. To address medical ethical issues, we must change our focus from articulating principles and rules that apply to individual cases to devising institutional structures that can ensure ethical behavior. (Emanuel, 1995, p. 338)

Similarly, Kevin Wm. Wildes, S.J. has argued that bioethicists must focus not only individuals but on the institutions and the web of relationships and obligations involved in health care. He characterizes a bioethics that looks only at individuals and ignores institutions as inadequate (Wildes, 1997, p. 413).

The field of organizational ethics has for some time addressed the moral obligations of health care organizations. Yet the issues of precisely how we are to understand such institutions as moral agents, what moral obligations we may justifiably attribute to them, and how we are to evaluate them remain the subject of disagreement.[1] I argue that to understand institutions as morally responsible we must understand them as having integrity because without an understanding of moral integrity we cannot understand agents' actions as having a purpose and we cannot understand agents as having moral obligations. Therefore, organizational ethics needs a robust concept of institutional integrity to understand and appreciate institutional moral responsibility.

Ana Smith Iltis (ed.), Institutional Integrity in Health Care, 1-6.

The importance of institutional integrity is underscored by the fact that we face the task of morally evaluating health care organizations in a society in which we lack a single, thick understanding of morality. The circumstance of moral pluralism complicates efforts to evaluate or assess health care organizations because we do not have a shared, robust concept of the right and the good that can guide our evaluations. It is because of this post-modern reality that we must turn to the concept of integrity. Integrity makes it possible to evaluate the extent to which health care organizations live up to their obligations regardless of what those obligations are. There are, in fact, two different kinds of obligations under consideration in this volume. Some obligations, it is argued, are borne by all health care organizations regardless of their moral commitments or identities. Because they are health care organizations, they are obligated in particular ways. Other obligations, however, may not be justifiably attributed to all health care organizations. These are the obligations grounded in an institution's moral commitments, and the ramifications of this are especially poignant in discussions of religious health care institutions.

A number of authors in this volume (see Werhane, Logue and Wear, and Reiser) focus on the obligations they hold are borne by all health care organizations in virtue of being health care organizations. Institutions of integrity, in those cases, are the ones that live up to the obligations established for them. But those obligations we might attribute to health care organizations in general do not give a full vision of integrity for health care organizations with robust moral commitments or thick moral characters. Religious health care institutions, for example, must do more than satisfy certain basic commitments attributable (or allegedly attributable) to all health care organizations. They must fulfill these responsibilities associated with their religious identities. Christopher Tollefsen and Duane Covrig attend to issues of integrity in religious health care institutions.

The collection of essays begins with Peter French's discussion of the moral elements of the social world in which he justifies the very project of treating and evaluating health care organizations as moral agents and bearers of moral responsibilities. In 'Inference Gaps in the Moral Assessment and the Moral Agency of Health Care Organizations,' French argues that the moral elements of the social world are varied and include (1) what French calls institutions, which are abstract entities such as the institution of slavery or the institution of medicine. French distinguishes (abstract) institutions from (2) organizations, which are the embodiments of institutions. This distinction is not operative in the remainder of the essays in the volume. (Other authors use the terms 'institution' and 'organization' interchangeably because they are not engaging in the same analysis French pursues.) Finally, (3) individuals are part of the moral element of the social world. The various elements are subject to moral analysis, and organizations such as health care organizations and individuals in particular have moral obligations. However, as French uses the "Jefferson Paradox" to illustrate, what we learn about the morality or immorality of one element of the social world does not necessarily tell us what another element ought to do or not do, or whether another element of the social world ought to exist or not. Each element must be assessed as such, and we should be careful not to assume a linear relationship between the various moral elements of the social world.

One implication of French's analysis is that we must be able to assess organizations, including health care organizations, and determine the extent to which they satisfy their moral obligations. The appropriate mechanism for doing this in a morally pluralistic society such as ours is through integrity. Hence, the focus of this volume is on understanding the integrity of health care institutions or organizations.

The second and third essays in the volume describe the reality of moral pluralism and the implications it has for understanding moral responsibility in general and moral integrity in particular. In 'Institutional Integrity in Health Care: Tony Soprano and Family Values,' Kevin Wm. Wildes, S.J. argues that health care organizations, like individuals, have different understandings of morality such that what one takes to be morally good or obligatory another may find morally, morally, or supererogatory. Integrity, he says, "is more than being consistent; it reflects a ranking of the goals, values, and projects that a person may wish to pursue" (Wildes, 2003, p. 33). As such, it is a second order virtue, which means that it cannot refer to an evaluation of an agent as good or bad in universal terms. We can only know whether an agent has integrity in light of her values, goals, and commitments. The moral assessment of those values, goals, and commitments is a separate project, and one that is complicated in a morally pluralistic society.

Similarly, Ronald C. Arnett and Janie Harden Fritz reflect on the reality of moral pluralism and its implications for understanding health care organizations. In 'Sustaining Institutional Ethos and Integrity: Management in a Postmodern Moment,' Arnett and Fritz develop an account of integrity in the post-modern moment Wildes describes. They take the reality of moral pluralism as a given and consider what integrity looks like in the organizational setting in the face of pluralism. Their account, like several others, applies to all organizations, whether they have thick or thin, religious or secular moral commitments. Organizations of integrity, Arnett and Fritz suggest, are those that maintain "a story or narrative that identifies [their] core values (mission)…supported by…social practices that give life to those values" (Arnett and Fritz, 2003, p. 41). Integrity is not necessarily intrinsically valuable on their account, but failures of integrity result in organizational failures. Incoherence between an institution's claims and actions generates cynicism and mistrust among employees, patients and patients' families. Among employees, this can lead to decreased commitment to the institution and to their jobs, which can lead an organization to fail to provide services as promised. To avoid this, managers must implement a narrative or story, bring people together into the organizational narrative, and show others why and how to be part of the story. In post-modernity there is no single story that holds individuals together; organizations must create a story grounded in their mission and then encourage others to be part of the story. Organizations with coherent narratives are organizations of integrity.

Wildes as well as Arnett and Fritz identify the complexities post-modern moral pluralism brings to efforts to morally assess health care organizations. Patricia Werhane, Gerald Logue and Stephen Wear, and Stanley Reiser, focus on the requirements health care organizations must meet to have integrity regardless of whether they possess any robust moral commitments. In 'Business Ethics, Organization Ethics and Systems Ethics for Health Care,' Werhane approaches the issue of integrity by looking at the health care system and the ways in which failures to work together result in harms. Institutions of integrity would be those that evidenced systems thinking and took into account the interests of all stakeholders.

While Werhane reflects on health care organizations as unique entities with distinct obligations, Logue and Wear treat healthcare organizations as having the same obligations individual health care professionals have. These obligations include the obligation to steward resources carefully and to respect patients' right to informed consent. They argue in 'The Health Care Institution/Patient Relationship' that as health care providers, health care organizations are subject to the same obligations we traditionally have attributed to individual health care professionals. It is more important than ever to take seriously the obligations of health care organizations, Logue and Wear argue, because today patients are more likely to have an ongoing relationship with a health care organization than with a particular physician. As such, the health care organization-patient relationship is of primary concern to health care ethics. Institutions of integrity are those that develop measures to fulfill their obligations by building a culture that attend to the reality of patients' rights and needs. Such institutions will also routinely assess themselves and evaluate the extent to which they fulfill their obligations to patients.

The chapter by Reiser, 'Creating an Institutional Ethical Identity,' addresses the minimum conditions necessary for a health care institution, regardless of its moral commitments, to be an organization of integrity. He argues that the health care ethics literature has focused to a large extent on the virtues of physicians and nurses and on the obligations of health care professionals. The notion of institutional virtue and moral character has not been sufficiently discussed. To begin to develop a notion of organizational virtue and organizational ethics, he says, we must be able to articulate what it is for a health care institution to have an ethical identity. Reiser identifies seven hallmarks of ethical identity for health care organizations. The points address organizations' relations both internally (e.g., with staff, students, and patients) and externally (e.g., with the public and future generations). According to Reiser's analysis, it would seem that for a health care organization to be one of integrity, it must, at a minimum, comply with these obligations.

The final two chapters in this volume bring forth again the reality of post-modernity and the moral pluralism that characterizes it. Although many health care organizations are secular in nature or have only modest religious commitments, others have robust moral and religious characters. Christopher Tollefsen and Duane Covrig address the integrity of health care organizations with strong religious identities. The integrity of such organizations cannot be understood in merely secular terms, and such organizations have moral obligations that we do not recognize as applicable to all health care institutions. In 'Institutional Integrity' Tollefsen argues that some institutions are "weak" in the sense that they do not have robust moral commitments. Integrity for such institutions is different from what is required of those with robust moral commitments, such as Roman Catholic hospitals and universities. The latter are what Tollefsen calls "strong" institutions, and to have integrity, Roman Catholic institutions must do much more than express concern for various stakeholders and take systems thinking seriously, uphold the rights of patients (such as the right to informed consent), and respect the internal and external communities that constitute and surround them. To have integrity, Roman Catholic institutions must live out their moral identities as Roman Catholic entities. This requires a different level of commitment from the individuals associated with them.

Covrig also addresses the issue of integrity for religious institutions with a focus on the Seventh-day Adventist tradition. In 'Institutional Integrity through Periods of Significant Change: Loma Linda University's 100 Year Struggle with Organizational Identity,' he discusses the history of Loma Linda University to show how individuals associated with the institution, including its top leaders, have struggled over the years to understand what its core identity and commitments are and what they require. His analysis does not reflect only the concerns of a Seventh-day Adventist institution, however. In his discussion of the difficulties Loma Linda University has faced with regard to maintaining its identity over time, Covrig points to one of the great difficulties in understanding institutional integrity. How ought an institution to maintain its identity, live out its commitments, and fulfill its obligations in a changing world? What compromises are permissible in the attempt to survive? In a changing health care climate in which institutional mergers have become common, these challenges are familiar to many, not only to those in the Seventh-day Adventist tradition.

The chapters in this volume bring together complementary visions of organizational integrity. The initial chapters suggest ways in which we might understand health care organizations as bearers of diverse moral obligations, and the latter chapters address the integrity of secular and then religious health care institutions. Despite the different approaches to understanding integrity, we can draw from the chapters on secular and religious institutions the important insight that institutional integrity is intimately connected to institutional identity. Some authors hold that all health care organizations have moral obligations grounded in their identities as healthcare organizations and thus the standards for integrity are universal among health care organizations. Others recognize that some organizations have identities that encompass more than merely being health care organizations because, for example, they have religious commitments. These organizations must satisfy additional obligations to possess integrity. Nevertheless, all the authors treat organizational identity as normative and thus as defining the requirements organizations must meet to have integrity. Together the chapters in this volume provide a significant contribution to the field of organizational ethics by providing insight into how we are to understand health care organizations as being morally responsible and how we are to morally evaluate them.

Saint. Louis University
Saint Louis, Missouri, USA

NOTE

1. The issue of whether institutions can be moral agents has been discussed in the examination of the status of states and corporations. To be moral agents, such groups would have to have an identity that is not fully reducible to their constituent individual parts. Thus, nominalists, such as Lon Fuller, and methodological individualists, such as J.W.N. Watkins deny altogether the possibility of institutional moral agency. Fuller (1967) argues that corporations (and other comparable legal structures) are created fictions accepted for convenience. Watkins holds that social phenomena "should be explained by being deduced from (a) principles governing the behavior of participating individuals and (b) descriptions of their situations" (1973, p. 88). Those who hold that social phenomena are reducible fully to individuals cannot attribute moral agency to groups, for they see that agency as reducible fully to individuals.

Others have presented analyses of social phenomena that oppose the nominalist and methodological individualist positions. Often these analyses have recognized the importance of individuals in constituting social phenomena – without the individuals associated with them, there would be no social phenomena. But they have argued that the identities of (certain types of) social phenomena cannot be reduced fully to those constituent individuals. According to Plato's organic theory of the state, the state is a whole made up of parts, only when all the parts are functioning properly can the whole be whole, and the whole cannot be reduced to its parts. The city has its own character, but this character depends upon the parts: if the city is unhealthy it is because (some of) the parts are not functioning properly and the city is healthy when the parts function properly. Thus for Plato the state is a person – a whole – whose identity depends on but is not reducible fully to its parts. In the contemporary literature on corporate moral responsibility we also find a range of views supporting the claim that corporations can have an identity not fully reducible to the individuals constituting them and that they can have some degree of moral responsibility. Patricia Werhane argues that corporations are moral agents whose identity cannot be reduced fully to the individuals constituting them. We should recognize corporate identity as dependent on but not fully reducible to the identities of the individual associated with the corporation. Moral responsibility is ascribed both to corporations and individuals; not all actions can be redistributed to individuals, she argues, and both individuals *and* corporations bear moral responsibilities: "Corporate moral agency is derived from individual moral agency, but it is neither identical nor reducible to it" (Werhane, 1989, pp. 821-822). Peter French argues that intentions and responsibilities can be attributed to corporations when they have a Corporate Internal Decision (CID) structure: "The CID structure is the personnel organization for the exercise of the corporation's power with respect to its ventures, and as such its primary function is to draw experience from various lines of the corporation into a decision-making and ratification process" (1979c/1977, p.177). As such, "corporations should be treated as full-fledged moral persons and hence that they can have whatever privileges, rights, and duties as are, in the normal course of affairs, accorded to moral persons" (1979c/1977, p. 176). French amends his terminology in 1995 and calls corporations 'moral actors' because of confusion the terminology of 'moral persons' caused (1995, p. 10). His main argument, however, remains unchanged. Corporations have an identity that is not reducible fully to the individuals associated with them and as such are bearers of moral responsibility.

BIBLIOGRAPHY

Emanuel, E. J. (1995). 'Medical ethic in the era of managed care: The need for institutional structures instead of principles for individual cases,' *The Journal of Clinical Ethics, 6(4)*, 335-338.
French. P. (1979c/1977). 'Corporate moral agency,' in Tom L. Beauchamp and Norman Bowie (eds.), *Ethical Theory and Business*, Englewood Cliffs, New Jersey: Prentice-Hall, pp. 175-186. Originally presented at the Ethics and Economics Conference, University of Delaware, November 11, 1977.
French, Peter A. (1995). 'Ethics and agency theory,' *Business Ethics Quarterly* 5(3), 621-627.
Fuller, L. (1967). *Legal Fictions*. Stanford: Stanford University Press.
Watkins, J.W.N. (1973).'Ideal types and historical explanation,' in Alan Ryan (ed.), *The Philosophy of Social Explanation*, London: Oxford University Press, pp. 82-104.
Werhane, Patricia H. (1989). 'Corporate and individual moral responsibility: A reply to Jan Garrett,' *Journal of Business Ethics* 8, 821-822.
Wildes, K.W., (1997). 'Institutional identity, integrity, and conscience,' *Kennedy Institute of Ethics Journal, 7(4)*, 413-419.

PETER A. FRENCH

INFERENCE GAPS IN MORAL ASSESSMENT AND THE MORAL AGENCY OF HEALTH CARE ORGANIZATIONS

1. THE "JEFFERSON PARADOX"

In the rough draft of the *Declaration of Independence* Thomas Jefferson wrote in justifying revolution against the English King:

> He has waged cruel war against human nature itself, violating its most sacred rights of life and liberty in the persons of a distant people who never offended him, captivating and carrying them into slavery in another hemisphere, or to incur miserable death in their transportation thither. This piratical warfare, the opprobrium of infidel powers, is the warfare of the CHRISTIAN king of Great Britain determined to keep open a market where MEN should be bought and sold, he has prostituted his negative for suppressing every legislative attempt to prohibit or to restrain this execrable commerce: and that this assemblage of horrors might want no fact of distinguished die, he is now exciting those very people to rise in arms among us, and to purchase that liberty of which he has deprived them by murdering the people on whom he also obtruded them; thus paying off former crimes committed against the liberties of one people, with crimes which he urges them to commit against the lives of another. (Jefferson, 1950, p. 426)

In his *Notes on the State of Virginia*, he writes about the future of slavery in America:

> Indeed I tremble for my country when I reflect that God is just: that his justice cannot sleep for ever: that considering numbers, nature and natural means only, a revolution of the wheel of fortune, an exchange of situation is among possible events: that it may become probable by supernatural interference! The almighty has no attribute which can take side with us in such a contest. - But it is impossible to be temperate and to pursue this subject through the various considerations of policy, of morals, of history natural and civil. We must be contented to hope they will force their way into every one's mind. I think a change already perceptible, since the origin of the present revolution. The spirit of the master is abating, that of the slave rising from the dust, his condition mollifying, the way I hope preparing, under the auspices of heaven, for a total emancipation, and that this is disposed, in the order of events, to be with the consent of the masters, rather than by their extirpation. (Jefferson, 1954, p. 163)

Jefferson acknowledges that slavery is a moral wrong, that the institution is at odds with the principles he so eloquently laid down in the *Declaration* and with respect to which the American Revolution was motivated. He proposed legislation to abolish the slave trade and the extension of slavery into the western territories,

Ana Smith Iltis (ed.), Institutional Integrity in Health Care,7-28.
© *2003 Kluwer Academic Publishers. Printed in the Netherlands.*

legislation that failed by a single vote. But throughout his entire adult life Jefferson owned hundreds of slaves on his two plantations, and he freed only a very few and those only from the Hemmings family. There appears to be little doubt that Jefferson was a racist, as a sampling of comments he wrote in the *Notes on the State of Virginia* attests: "I advance it therefore as a suspicion only, that the blacks, whether originally a distinct race, or made distinct by time and circumstances, are inferior to the whites in the endowments both of body and mind" (1954, p. 141). He also states: "...their inferiority is not the effect merely of their condition of life." "It appears to me that in memory they are equal to the whites; in reason much inferior, as I think one could scarcely be found capable of tracing and comprehending the investigations of Euclid; and that in imagination they are dull, tasteless, and anomalous" (Jefferson, 1954, p. 139). He, nonetheless, identified the enslavement of blacks as immoral, though he regarded blacks as unfit by nature to participate in the American social experiment.

The "Jefferson Paradox", as I use the term, does not, however, refer to his support for emancipation while he held and championed outrageously racist doctrines. Odd as it may sound to us, for Jefferson those were not incompatible positions. That is because he believed that blacks were, in fact, the equal of whites in the moral sense and that they thereby had the same natural rights as all humans and deserved to be free. However, his doctrine of virtue was a more complex matter and he found blacks sorely lacking in that regard. For Jefferson, virtue consisted in a union of memory, reason, sentiment, moral sense, and imagination. It melded aesthetic and moral elements so that the virtuous person not only must have a well-developed sense of right and wrong, but a sensibility to beauty, an appreciation of nature, and a cultured imagination. Jefferson was convinced, despite counterexamples in his own experience (for example, Benjamin Banneker with whom he had corresponded), that blacks were woefully lacking in the faculties of reason, imagination, and aesthetic sensitivity, and shortcomings in those areas precluded them from the achievement of true virtue. They were, he believed, not only unsuited for integration into the white society of America, but they were a positive threat to the creation and maintenance of public and private virtue in the newly independent white population. For Jefferson, emancipation must be followed by colonization of the former slaves at some place remote from the shores of America, as he wrote to James Monroe: "Could we procure lands beyond the limits of the US to form a receptacle for these people?"[1]

The "Jefferson Paradox" that interests me, though in his case undoubtedly related to his racism, is not directly rooted in that set of his beliefs. Indeed, it is the less complex one that leaps out at anyone with but a modicum of knowledge about Jefferson: the author of the *Declaration of Independence* who also fervently believed that the institution of slavery was immoral, bought, owned, and sold slaves throughout his life. His standard of living was utterly dependent on his participation in an institution he believed to be morally wrong. Yet, he never condemns himself for his active participation. It is tempting to think that Jefferson is the epitome of hypocrisy, and we would think that because we tend to think that from a slave owner's assenting to the judgment that "the institution of slavery is immoral and should be abolished" some course or courses of action follow on the part of that slave owner with respect to his/her slaves: that one should free them.

The argument looks like this:

> [A]   (1) The institution of slavery is immoral and should be abolished.
>   (2) I own slaves.
> Therefore (3) I should free my slaves.

Such a conclusion, from a practical point of view, might be said to follow from the premises if attached to it is a *ceteris paribus* clause: I should free my slaves, other things being equal, or something of that sort. Or, it might be read that I have a *prima facie* moral obligation to free my slaves. That obligation, being *prima facie*, of course, may be over-ridden by more important moral concerns, and in Jefferson's case, he believed that such concerns existed. Ethically speaking, of course, such conclusions drawn from those premises are a cheat, for the very reason that things in such cases are seldom equal or morally insignificant. I want to argue that it is not at all clear what conclusion(s), morally speaking, regarding my behavior as a slave-owner follows from those premises. It seems to me to be anything but clear that Jefferson morally ought to have freed his slaves, even though both premises are true and he would have agreed that they were so.

Marcus Singer, in a paper that has not received the attention it deserves, argues "no moral judgment of an institution-constituted action follows from a moral judgment of the related institution" (1993, p. 239). By "an institution-constituted action" Singer means an action whose description as an action of that sort depends on the existence of an institution or social practice for its sense. Simply, in the absence of the institution no one could perform the action. Owning slaves and freeing them are examples of such actions with respect to the institution of slavery. Singer's point is that from the negative moral assessment of the institution all that could follow deductively is that we have a moral problem with the institution. What morally ought to be done by an individual engaged in actions constituted by the institution is not inferable from the assessment. If Singer is right, then, though I concur wholeheartedly with [A] (1) and acknowledge that [A] (2) is true, it would not be illogical for me to decide, even having adopted the moral point of view, that I should not free my slaves.

Borrowing somewhat from Singer, what I morally ought to do with my slaves is dependent on crucial factors other than the moral assessment that the institution of slavery is immoral and should be abolished. Which is not to say that the moral assessment regarding the institution is utterly irrelevant with respect to what I ought to do with my slaves. Jefferson seemed to have unromantically examined factors of the relevant sort. He lived in a slave-owning society, one in which slavery was firmly established at the root of its economic system, a society in which law and custom supported it. He could have reasoned that if I were to free my slaves, the law might allow someone else with a reputation for extreme sadistic cruelty towards blacks to capture and enslave them. It is possible, even likely, that if I were to free my slaves, they would find it impossible to make a living, impossible to survive. It is likely that I might be condemning them to a life much worse than I provide on my plantations because no provisions are made in this society for blacks who are freed or ex-slaves to participate in the community in a meaningful and productive or even basic way other than as slaves. To free them would be to turn them out into, at best, a wilderness with the barest chance of survival or to place them in extreme danger of loss of life, limb, or whatever modicum of freedom I can now afford them. In 1820 Jefferson, in fact, made what we would regard as a paternalistic argument along similar lines. He wrote: "Nothing would induce me to put my Negroes out of my protection." He maintained that freeing those who have spent their lives in slavery would be "like abandoning children." All of these "considerations" are, of course, disputable suppositions and someone could reasonably maintain that even if freeing one's slaves resulted in the worst of these outcomes, perhaps an hour or two of free choice in their actions before they met a worse fate than under one's "protection," one still morally ought to free one's slaves. But the very fact that these possible outcomes of emancipation are clearly considerable from the moral point of view and it would be at best morally callous to ignore them when one decides what to do with one's slaves, suggest that the individual action decision is not directly inferable from the premises of [A]. In any event, I am not really interested in slavery or Jefferson's mindset. I am concerned (with Singer) to note that there is an inference gap of sizable proportions between the premises of [A] and any conclusion about what a slave-owner (like Jefferson) morally ought to do.

Another inference gap problem should be noticed in the premise [A] (1). That premise rests on the generalized claim that practices or institutions that are immoral ought to be abolished. In the case of slavery that is indisputable, or at least it is in the case of forced slavery (contractual slavery may be another matter, but not one I want to explore), but that is because slavery is a special case among institutions. It may not be the case that most other institutions in a society that may rightly be judged to be immoral in very serious ways ought to be abolished. In this I depart from Singer's account. For example, if I were to convince you that our institution of criminal justice is seriously flawed in very fundamental ways from the moral point of view (because it is unfair to those from lower economic classes), it does not follow that you must endorse abolishing the institution. You must, on pain of inconsistency, be willing to endorse doing something to correct its faults, its moral deficiencies, that is, reforming it, fixing it. But reform is not abolition. The same could probably be said of the institution of medicine (understood, for example, in terms of health care delivery across the population). The reason we cannot reform slavery to satisfy our principles of justice and fairness, particularly those that protect human dignity, respect, and worth, is that were we to do so, the resulting institution would not be slavery. So, though it does follow, when the moral status of slavery is unpacked, from "slavery is an immoral institution" to "it must be abolished," such an inference cannot be generalized to all institutions that are rightly assessed to be immoral or immoral with respect to some of their fundamental elements or to some, perhaps significant, degree. What does follow is that something corrective, perhaps even radical, must be done about the institution for it to pass moral muster.

However, nothing specific about what individuals engaged in the institution, some of whose actions are institution-constituted, practitioners, morally ought to do with respect to exactly those actions seems to follow from the negative moral assessment of the institution in and of itself.

Jefferson seems to have realized this. He wrote and voted in support of the abolition of slavery, recognizing that the institution could not be reformed, but he also seemed convinced that its abolition was not a matter of individual action, including his, *qua* slave-owner. It was a collective task, a communal moral obligation, and one, he believed, that the fledging country of the United States of America was incapable of undertaking in his lifetime.

I do not know whether Jefferson was a hypocrite when he wrote about the evils of slavery. It is virtually impossible to read him as anything but a racist. I have no real interest in such matters in Jeffersonian scholarship. My intent is just to draw what I think is an important lesson from what I have called the "Jefferson Paradox:" it is a mistake to think that negative moral assessments of institutions entail that those who perform actions that are constituted by those institutions ought or ought not to be performing them. The inference gap exists in both directions, from the moral assessment of the individual's institutionally constituted actions to the moral status of the institution or the system of rules that constitute it and vice versa. For example, it does not follow that because a certain married couple morally ought not to get a divorce that divorce is immoral, ought to be prohibited, that an institution of marriage that permits divorce ought to be abolished or significantly reformed. The inference gaps expose a very important structural aspect of the moral world. They reveal, I will suggest, that the elements of the moral world, though related, are distinct and require separate moral attention when it comes to moral assessment.

Institutions are abstractions. They are, as Singer, leaning heavily on Rawls, describes them, "complex(es) of rules defining rights and duties, roles, functions, privileges, immunities, responsibilities, and services" (Singer, 1993, p. 228). To use a well-worn example in the philosophical literature, baseball is an institution that evidences all of those aspects. Most, though not all, of the institutions in our social system are concretely embodied in what I will call organizations. The institution of baseball is concretely embodied in the Major League Baseball organization, and also in the Little League organization, and yet again in thousands of other organizations ranging from the amateur leagues to the semi-professionals, etc. And, of course, individuals play baseball by performing institutionally constituted actions. There is baseball, Major League Baseball, Incorporated, and Randy Johnson pitching. There was slavery, the Monticello Plantation, and Thomas Jefferson owning, emancipating, buying, and selling slaves.

As mentioned above, not all institutions have concrete organizational embodiments. For an example, consider promising. Promising, however, is a prerequisite institution for other institutions that do have organizational embodiments even if only rudimentary ones, e.g., marriage. There may be multiple linkages between institutions on which an organizational embodiment of one or more than one is or are based. There are also institutionally constituted actions that persons can perform in the absence of any specific organization: my promising to meet you for lunch tomorrow.

Rawls maintains that the principles of justice for institutions must not be confused with moral principles that apply to individuals. "These two kinds of principles apply to different subjects and must be discussed separately" (Rawls, 1971, pp. 54-55). He adds that institutions as systems of rules are abstract objects, but that they are also "realized." By that I take him to mean that they are actually practiced in social life. Promising may be understood as an abstraction that will be just if certain principles govern it, but it is the institution as realized in the behavior of people in a society that is just if it is practiced in accord with certain principles. I want to distinguish, where Rawls is only suggestive, between the realized institution and an institution's concrete embodiment in an organization. On the account I am favoring, the institution of promising can be seen as an abstract object and as realized in the occasions in which people promise each other different sorts of things, e.g., to meet for lunch, the loan of a book, the repayment of a debt, etc. But, as noted above, promising per se does not have a concrete embodiment in an organization, though it plays a significant role in many organizations. I would extend Rawls' point that we should not confuse the moral principles and judgments made with respect to institutions with those relevant to individuals practicing within those institutions to organizations that embody institutions, or through which many institutions become effective in social life.

This may be somewhat harder to accept than the distinction that Rawls admits between the principles that govern institutions and those that ought to govern individual behavior within those institutions. After all, organizations, on my account, are to be understood as concretely embodying institutions. I would have us think of an institution as the coherent organizing of a certain set of norms and that the institution is realized when a group of people act in ways that they understand to accord with those norms. An organization is a social structure that concretizes, systematizes, arranges, administers, and typically polices the roles and activities of persons that are made possible by the rules and norms of the institution. So some institutions are realized through embodying organizations or even fleets of organizations.

Imagine that the realized institution of medicine (through a number of associated organizations) falls significantly short of achieving what justice requires with regard to the availability of medical care for all members of the society. (Call that judgment [S].) (I have no interest in trying to make an argument to support this claim, though I suspect it is the case.) Then think of those tightly or loosely associated organizations that concretely embody the institution of medicine in our society, such as hospitals, clinics, physician groups, HMO's, etc. What judgment follows from the negative assessment of the justice of the institution with respect to the institution-constituted actions of a particular hospital or clinic? Does the judgment that private hospital H ought to adopt a policy of providing needed health care free of charge to indigents that happen on H's doorstep follow from [S]? I do not think that it does. It might be persuasively argued that H ought to adopt some set of operating policies that respond to [S], that we would hold H morally deficient if it did not do so. Whether or not those would include providing free health care to indigents is not clear. On the other hand, all that might follow from [S], morally speaking, with respect to H is that H ought to work, in conjunction with other concrete embodiments of the institution of medicine, public and private, towards the reform of the realized institution while continuing to conduct business as usual: rejecting service to indigents, directing them to public hospitals or clinics.

With respect to institutions, their concrete embodiments, and those who practice within them (perform institution-constituted actions specific to those institutions) it makes sense to worry about the appropriate moral evaluations and assessments made relative to each element in large measure independent of the others because there exist significantly wide inference gaps between moral assessments made about institutions, those made with respect to their concrete embodiments, and those made about the actions of individual practitioners. That point seems virtually incontestable regarding institutions and individual practitioners, as noted by Singer and Rawls, but neither has anything to say about organizations. In Rawls' case that is no surprise because he tends to adopt a methodological individualist's conception of organizations. On such an account organizations are just collections of individuals and moral and other judgments made about them are supposed to reduce without remainder to like judgments about the individuals who are organized within them. I have argued against that view for about three decades. On my view organizations, such as hospitals, clinics, etc. should be understood as moral agents in their own right and therefore bearers of moral responsibility. Institutions can be morally assessed, but they cannot be held morally responsible, though their concrete embodiments that qualify as moral agents can and should be held accountable for their moral defects and failings.

2. ORGANIZATIONS AS MORAL AGENTS

In the history of ethics one can find philosophers who think that the basic or primary point of ethics should be to make people good and those, like myself, who believe that minimizing the undeserved harm inflicted on people and the environment should be the most important goal of ethics. These are not, to be sure, incompatible goals. The matter is one of focus or emphasis, but emphasis does make a considerable difference. Certainly, if we ever succeeded in making everyone good, we would go a long way towards eradicating the undeserved harm that we perpetrate on each other. A long way, but not all the way![2] If the focus of ethics is shifted from making people good to minimizing undeserved harm, our foremost concerns will be with identifying harm-causers and effectively dealing with them. Which is not to say that prevention will not also be a major concern. My view has long been that corporately organized entities, organizations of a certain sort, are more capable of causing undeserved harm across significant portions of the population and the environment than are individual human beings. And that, in no small measure, is because they embody our institutions. If an ethical theory systematically cannot address the organization offender *qua* organization offender then it will be impotent with respect to minimizing a great deal of undeserved harm, and, I believe, a failure in ethics.

Opposed to treating organizations as morally assessable agents in and of themselves is the long-standing philosophical conceit in ethics that favors methodological individualist accounts rather than what Margaret Gilbert (1989 and 2000) calls "holistic" accounts of groups and organizations. Methodological individualism is the theory that the behavior or any aspect thereof, such as intention, of an organization is always reducible to the behavior or intentions of the individuals who make up the organization (without remainder). I strongly suspect that it may be impossible to coherently state the theory of methodological individualism. In any event, the theory simply does not adequately explain a significant number of social facts concerning organizational behavior and the intentions and commitments of plural subjects and it crucially ignores the role of organizational culture and structure in decision-making, two elements that mark the identity of any organization and that are essential in reidentifying an organization over time.

In my work on corporate or organizational moral agency I defended three positions: (1) corporations or corporately structured organizations exhibit intentionality that is not reducible to the intentions of the individual members of the organizations, (2) organizational intentions can be rational, and (3) organizations can alter their intentions and patterns of behavior for any number of reasons. The keystone element in my picture of such organizations, of course, is my claim that it makes sense to understand some organizational behavior in terms of organizational intentionality.

In my earlier work I held the rather traditional view that having intentionally done something is essential to being held morally responsible for its occurrence. I no longer think that such a strict intentionality condition is defensible. People and organizations often do considerable undeserved harm when not intending their actions under the specific description of which the causing of that harm was an explicit element. In such cases, of course, the agent does not act without intention. But the agent does not intend the action under the description that would have been required for moral responsibility on the strict intentionality condition. Some of Aristotle's ignorance cases might fall into this category, but I suspect that the majority of cases are those in which, for whatever reasons, the agent(s) are factually but not logically incapable of seeing the circumstances in which they act or the outcomes of their actions under descriptions that describe them as undeserved harm-causing (French, 2001b). I am no longer persuaded that the intention to do the deed under the relevant description is as important to moral responsibility as the fact that undeserved harm was done and that morality must not allow such actions to pass without condemnation and punishment. Nonetheless, intentionality still sits at the core of the issue of moral agency and so organizational agency and organizational moral responsibility. Simply, I maintain that something must be a moral agent, that is an entity that is functionally capable of acting intentionally, in order to be held morally responsible for its chosen and unchosen harm-causing (French, 2001a, pp. 194-205). That is not, I think, a controversial position.

My intent when I first raised the issue of organizational moral agency was to provide a structure that would allow us to understand how describing an action as "organization-intentional" makes sense and I worked out the Corporate Internal Decision Structure (CID Structure) theory to serve that purpose (French 1979 and 1984). At the core of my earlier view was the widely held position that intentionality should be understood in terms of a desire/belief complex. A number of philosophers, including Alvin Goldman (1970), Elizabeth Anscombe (1963), Donald Davidson (1980), and Robert Audi (1973), in their analyses of intentionality and intending, focus exclusively on intention as it seems to appear in actions rather than, in Michael Bratman's terms, "in the state of intending to act" (Bratman, 1987. p. 5). Because they do that, they seem compelled to believe that "what makes it true that an action was performed intentionally, or with a certain intention, are just facts about the relation of that action to what the agent desires and what the agent believes" (Bratman, 1987, p. 6). For some, like Davidson in his early papers on the subject, that relation is made out to be a causal one. Others, such as Audi, argue that there is no state of intending because intentions with respect to future activities always reduce to appropriate sets of desires and beliefs. In other words, if I say, "I intend to go to Colorado," I am saying no more than that I desire to make the trip and believe that I can make it.

If intention is understood on the popular desire/belief model it is natural to think that any talk of the intentions of organizations (and so organizations as moral agents) must be metaphorical or reducible to the intentions of a human (or humans) who has (or have) the requisite desires and beliefs. Organizations cannot, in any normal sense, desire and believe. Consequently, in my earlier accounts I redescribed desires and beliefs into organizational policy in order to match the model. Many objected that I had overly formalized the notions of desire and belief to fit the CID Structure approach I had created. I am now prepared to say that if intention must be understood on the desire/belief model, then organizations will fail to make it as intentional agents, and so the undeserved harm caused by organizations will be out of the reach of morality, except as it can be attributed to specific individuals that are intentional, that is desiring and believing, agents.

Adopting the desire/belief complex conception of intentionality, however, was a mistake that I have since corrected in my work on intentionality, organizational or individual. To intend to do something, I now believe, following the work of Bratman, is to plan to do it, and though there may be desires and beliefs provoking planning or related to planning, planning is not just or reducible to a desire/belief complex. If I intend to go to Colorado in June, then I plan to go to Colorado in June. Or, at least, I have made some plans to do so. It is not that I just desire to go there in June and have a belief that I can do so. In fact, I may intend to go even though I do not desire to go there. I would rather go to Ireland, but I agreed to attend a homeowners meeting in Colorado. I am committed, resolved, to doing it. That is what it is to intend to do it. To say that some entity acted intentionally is to say that his or hers or its actions were planned, or undertaken deliberately to accomplish a goal(s); they were schemed, designed, even premeditated by that entity. I may do little to indicate what intentions I have, what my plans are. I may put off buying the tickets and packing my clothes. You might not be able to tell from any of my present behavior that I am intending to go to Colorado in June. The reason for this is that I might now be doing any number of things that are compatible with my intention to go to Colorado in June, though they have nothing to do with a Colorado trip. On the other hand, I cannot be intending to go to Colorado in June if I book up that month with trips to Europe and Asia. So if I intend to go to Colorado in June, some things are excluded from my possible present activities. That is what it means to be committed to, to plan on, doing something.

In effect, my intention seems to have little to do with my current desires and beliefs. In fact, desires and beliefs are, at most, only tangentially involved. My plans and my commitments to those plans are at the heart of my intentions. Of course "plan" might be used in at least two ways, one to refer to a set of plans, as for example a blueprint, and secondly to talk of what one plans to do. "Here is the plan that I plan to bring to reality." It is the second sense that captures intentionality, the sense in which "plan" entails commitment. The rejection of the desire/belief complex model of intentionality also will remove most of the subjectivism from my earlier accounts of organizational intentionality.

Bratman writes: "Plans are not merely executed. They are formed, retained, combined, constrained by other plans, filled in, modified, reconsidered, and so on. Such processes ... are central to our understanding of ... intention" (Bratman, 1987, p. 7). Such an account helps to explain why it is senseless to ask someone whether he intentionally sat in a chair if, when he entered the room, he just sat down in the chair. To raise the question of intention suggests that he was up to something besides sitting down. Was it an act of protest? Austin is correct when he notes: "'I intend to' is, as it were, a sort of 'future tense' of the verb, 'to X.' It has a vector, committal effect like 'I promise to X,' and again like 'I promise to X,' it is one of the possible formulas for making explicit, on occasion, the force of 'I shall X' (namely, that it was a declaration and not, for example, a forecast or an undertaking)" (Austin, 1970, p. 279).

I am not saying that intentions never involve desires and beliefs. They may enter into the various planning stages. We should be surprised if they did not. Austin talks of the machinery of action involving such stages or departments as intelligence, planning, decision, appreciation, and resolve (Austin, 1970, pp. 193-194). "I plan..." might well express one's having worked through at least Austin's "departments" of planning, decision, and resolve.

The decision-making and action processes of organizations, as anyone who has studied them can attest, very closely mirror the planning model of intention developed by Bratman. Planning is indisputably a major organizational activity. What may be plausibly disputed, however, is whether organizations plan or just their managers plan and so only the managers can be said to intend and have commitments, etc. If such a view were adopted then organizational intention would be the shared intention of, usually senior, managers, typically speaking in the form "We intend to ..." and meaning "This is what the organization will do." On Bratman's account of shared intention the organization's managers' announcement, "We intend to do A" is shorthand for Manager X's "I intend that we do A" and Manager Y's "I intend that we do A", etc. where each manager's expression of individual intention is made with common knowledge of, in accordance with, and even because of the intentions (or what Bratman calls the "subplans") of each of the other managers. Bratman's account of this process is, as Gilbert has persuasively argued (Gilbert, 2000, p. 156), ultimately an individualistic construal of shared intention. I think it does not accord with the way planning, decision-making, and action occurs in corporately structured organizations.[3] It may, however, be on target with respect to the shared intentions of relatively unstructured groups, but even then I worry that it dismisses what Gilbert calls "commitment of the whole" (Gilbert, 2000, p. 158) in joint undertakings. In any event, that matter can be left for another occasion, while, for current purposes adopting the planning theory of intention and rejecting the desire/belief model.

I hope that a cursory mention of work I have previously done on Corporate Internal Decision Structures (CID Structures) will suffice to make the case for organizational intentionality on the planning theory. An essential feature of a corporately structured organization is that it has an established way by which it makes decisions and converts them into actions, a CID Structure. CID Structures have two elements crucial to our understanding of how intentionally acting organizations emerge at certain levels of the description of events: (1) an organizational flow chart that delineates stations and levels within the organization; and (2) rules that reveal how to recognize decisions that are organizational ones and not simply personal decisions of the humans who occupy the positions identified on the flow chart. These rules are typically embedded, whether explicitly or implicitly, in statements of organizational policy.

Its CID Structure is an arrangement of its personnel for the exercise of the organization's power with respect to its ventures and interests and, as such, a CID Structure's primary function is to draw various levels and positions within the organization into decision-making, ratification, and action processes. A CID Structure subordinates and synthesizes the intentions and actions of various human persons (and sometimes even the behavior of machines) into an organizational action. What I mean by that is that the CID Structure not only organizes the various human beings in the organization into a decision-making and acting entity, it makes it possible for us and those within the organization to describe what is happening as an organization's actions, plans, positions, etc. and not just as the actions of a specific manager or officer of the organization. In the absence of the structure, many of the activities of the humans and machines would be utterly unintelligible. To return to the baseball analogy, the actions of the pitcher would be unintelligible without the structure defined and created by the rules of the game. But further, it should be noted, some of the actions of the pitcher would still be unintelligible in the absence of knowledge of how the game is played, strategy, tradition, and custom internal to the game.

CID Structures generally are, and need to be, epistemically transparent, significantly differing from humans whose decision-making is rather opaque, even to the very humans making the decisions. Anyone with access to CID Structures should be able to discover everything about how they work. Hence, CID Structures can be confidently used as licenses of redescription to transform descriptions of certain events as the actions or the mere behavior of humans and/or machines into descriptions of actions of the organization. A CID Structure provides the means by which we gain access to a certain kind of intentional agent, namely organizations, at a different level of description than the one we typically use to describe the behavior of individual humans. And, as maintained in Part I, this is an ethically distinct level, with respect to inferring morally required or appropriate actions, from the level on which we morally assess human behavior and the institutional level as well. In effect, the attribution of organizational intentionality is referentially opaque with respect to other possible descriptions of the event in question, for example, as the intentional actions of a specific member of the organization.

I have elsewhere identified two sorts of rules: organizational rules and policy/procedure rules, in CID Structures (French, 1979). The organizational rules distinguish players, clarify their rank, and map out the interwoven lines of responsibility within the organization. They provide the grammar of organizational decision-making. Policy/procedure rules supply its logic. Every organization creates a general set of policies and procedures that, at least ideally, should be easily accessible to both its members and those with whom it interacts. When an action performed by a *bone fide* member of an organization is an implementation of its policy, and accords with its procedural rules, then it is proper to describe the act as done for organizational reasons or for organizational purposes, to advance organizational plans, and so as an intentional action of the organization.

The recent Arthur Andersen obstruction of justice case related to the Enron collapse may illustrate the point I am making. The government in prosecuting Andersen argued that its employees followed the policies and procedures found in the Andersen CID Structure when they destroyed accounting documents relative to Enron's accounting practices. Hence, the company and not just a few of its employees should be found guilty. Andersen defended itself by maintaining that the employees involved were acting on their own and not following corporate policy. That defense became harder to sustain as the prosecution piled up the evidence regarding the corporation's policies and decisions.

The plans of an organization might, perhaps often, differ from those that motivate the human persons who occupy positions in its CID Structure and whose bodily movements are necessary for the organization to act. Using its CID Structure, we can, however, describe the concerted behavior of those humans as organizational actions done with an organizational intention, to execute an organizational plan or as part of such a plan. Organization intent then, is dependent upon relatively transparent policies and plans that have their origins in the sociopsychology of a group of human beings. Organization intent might look like a tarnished, illegitimate offspring of human intent. If, however, we concentrate on the possible descriptions of events and acknowledge that there are distinctly organizational plans and policies that provide the reasons why organizations do the things they do, then we should not feel compelled to reduce statements about organizational actions to ones about the actions, reasons, plans, or interests of humans who happen to hold membership in the organization. We do not have to adopt Bratman's individualistic account of shared intention. Actually, I do not think that organizational intention is necessarily a version of shared intention, though organizations probably run smoother if at least their managerial team members have identified their plans with those adopted by and in the interest of their organizations (French, 1995). Recent events in the American corporate world, for-profit and not-for-profit, suggest that when that is not the case rather spectacular disasters can befall the organization. Think of Enron, Global Crossing, Adelphia, Worldcom, ImClone, and the United Way.

Organizations make plans, in fact, if anything, they are planning entities, designed to do just that. An organization's policies provide sufficient reasons and the other intentional elements needed to redescribe certain events as organizational actions, allowing, of course, that the member's actions are procedurally correct. The recognition of procedural rules in CID Structures is a bit easier than identifying policies, even though some procedures may be the result of common practice rather than official sanction. The policies of an organization seem to be inviolate. The basic ones generally are, indeed, they have to be for reasons of maintaining organizational identity. In that respect they are unlike policies adopted by individual humans. You could adopt a policy of honesty, but you may occasionally violate that policy by lying. When you lie, it is still you lying. If members act in ways that violate organizational policy, their acts are no longer organizational. But whether or not a policy is actually in place in an organization is dependent not on just what is on the written record of the organization, but on how members respond to apparent violations of it. So the sociology of the organization is often necessary in identifying its real policies.

To be a proper target of ethical evaluation, it seems to me indisputable, something must be capable of responding to what it learns about those with whom it interacts, as well as to ethical criticism. It must be able to responsively adjust its patterns of behavior and its policies and procedures to proactively head off predictable failures as well as prevent reoccurrences of disvalued outcomes. Such a responsive capacity, especially to moral criticism, seems to me to be crucial to anything being considered a moral agent. Organizational policies must be somewhat flexible so that organizations can respond to unexpected circumstances in ways that will further their interests, and fully realized CID Structures build in that capacity. In effect, organizations, if they are to survive for very long, must be capable of making rational non-programmed decisions directed to the satisfaction of their interests. Their CID Structures must encourage some reactive, responsive, discriminatory elements with respect to policies. Insofar as organizations demonstrate non-programmed responsive decision-making capacities, they would seem to qualify as moral agents in that regard, as well. It therefore should not be too troubling to talk about organizations as members of the moral community, to treat them as proper subjects of moral assessment and judgments and hold them morally responsible, even if those judgments do not reduce to similar judgments about individual human beings or if similar judgments about individual human beings who work within or are members of organizations are not directly inferable from them.

To summarize to this point: I have maintained that there is a substantially wide inference gap between moral judgments or assessments of institutions and like moral judgments or assessments of the organizations that are concrete embodiments of those institutions. I have also maintained that those organizations should be thought of as moral agents in their own right and not as mere collections of human individuals. Let me again refer to the Jefferson situation for illustrative purposes. We should grant that the institution of slavery is immoral and must be abolished whenever and wherever it exists. And we must admit that Monticello Plantation during the Jefferson years was a slave-based organization. From that we cannot directly infer that during the Jefferson years the Monticello Plantation should have been abolished. It is not unimaginable, morally speaking, that one might reach the conclusion Jefferson reached: that though the institution must be abolished, Monticello Plantation should not be abolished at that time. Of course, in the case of slavery, this becomes a very hard sell, because forced slavery is inherently immoral and the organizations that embody it surely are also so. That I think is undeniable and fuels our concern that Jefferson was a hypocrite, as well as a racist. If the institution is not one that is inherently immoral, but riddled with unfair or unjust practices, inferences to the operations of the specific organizations that embody it are more problematic and perhaps more clearly illustrate the inference gap problem. I think that is especially true in the case of the institution of medicine and the health care organizations that are its concrete embodiments in our society.

3. HEALTH CARE ORGANIZATIONS AS MORAL AGENTS

Many, if not most, of the organizations that concretely embody our institution of medicine are corporately structured, that is, they are constituted by CID Structures. Hospitals, whether or not they are for profit, mirror the corporate model in their managerial structures. The same can be said of clinics, HMO's, and the insurance companies that now virtually control the delivery of health care in the society. They all, more rather than less, satisfy the constitutive and regulative conditions of well-constructed CID Structures. Insofar as those structures are designed in the health care organizations for virtually the same purposes for which they exist in corporate entities in the business sector, of which strategic planning and evaluation, reassessment, and rational response are major elements, the health care organizations that dominate the concrete embodiments of the institution of medicine in our society are capable of organizational intentionality and qualify as proper subjects, *qua* organizations, of moral assessment and bearers of moral responsibility.

It would be redundant to repeat the discussion of Part II by making the general points specific to the health care organizations. On some occasions, probably frequently, they act organization-intentionally to perform a number of actions and as organizations they ought to be held morally responsible for those actions. They set admission policies and procedures, make contracts with physicians and other providers, negotiate payments with insurance carriers, HMOs, and patients, open and close clinics and various types of units, relocate hospital facilities, make deals with various government agencies, purchase equipment, upgrade facilities, report financial statements to various entities and individuals, etc.

Consider the following story that is based on an incident that occurred in the summer of 1990. Two men in their mid-thirties were attempting a climb to the summit of one of the "fourteeners" in Colorado. One of them, X, lost his footing on a narrow trail not far from the summit and fell 25 or 30 feet into a ravine. He was seriously injured and lay bleeding. The other climber, Y, was able to reach another party of climbers who eventually got word of the accident to a search and rescue unit and a helicopter was dispatched and X was airlifted to a relatively nearby well-equipped and staffed for-profit hospital. Y had to check the unconscious X into the hospital and Y was aware of the fact that X, who was between jobs, had allowed his health care insurance coverage to lapse. The hospital admitting clerk, C, informed Y that X must have insurance or the most the hospital would do in its emergency unit was try to stabilize X and then arrange for him to be transported via ambulance to a public hospital some 150 miles away. Y was convinced, and apparently with good reason given the extent of X's injuries, that X's life was at risk unless he received high-quality medical care immediately at the private hospital to which he had been flown. Y gave the admitting clerk his own name as X's name and produced his insurance card. X was given the needed quality care and survived. The hospital and transport bill came to approximately $50,000, and that amount was billed to Y's insurance company. Some months later the insurance company learned of what Y had done and they brought charges against Y and X for fraud. The two were convicted and sentenced to prison terms and required to make full restitution to the insurance company.

Admittedly, this is a dramatic incident and probably not one on which any sweeping generalizations regarding health care organizations or the system of health care delivery ought to be based. In any event, that is not my purpose in raising it. Suppose we, more or less, pass over Y's actions in this matter. At some level, probably a moral one, Y, rather than being condemned as a liar and defrauder, might be commended for acting in a way that he probably knew put himself in jeopardy in order to secure the needed medical care for X, his friend. He might be praised for taking the risk and in doing so, possibly, even probably, saving his friend's life. Good for Y, but I want to focus on the organization, the hospital, one of the organizations embodying our medical institution. I will not be concerned with another such concrete embodiment: the insurance company or what is called the "third-party payer." The hospital, in question, is corporately organized. It has a CID Structure that undoubtedly was designed to mirror those of most business corporations, and that it not just because it is a for-profit enterprise. Hospitals, regardless of their profit-making status, are managed structurally along the same lines as business corporations, as noted above.

C is not acting as an individual when she carries out her duties in the role of admitting clerk. She, as she probably told Y, was only following hospital policies and procedures and those policies and procedures are embedded in a number of documents that have been developed by the hospital's managers and endorsed by its board of directors, etc. according to the logic and grammar of the hospital's CID Structure. Probably the crucial ones regarding admitting patients are prominently posted in C's office for admitting clerks and would-be patients to read. In effect, then, the policies and procedures that C enforces were the result of a process of decision-making, planning, etc. that occurred in accord with the CID Structure of the organization. The establishment of those rules and procedures for admission was an organization-intentional action of the hospital, and, from an economic point of view, we should note, the adoption of a proof of ability to pay policy is a wise and rational decision for such a for-profit hospital to make. If the policy is as strict as the story suggests or as C presented it to Y, we might have some qualms about it from a moral perspective. Suppose that we do. Suppose that, morally speaking, we believe that the hospital's admission policies ought to allow that in extreme trauma cases like X's the hospital will admit the patient and provide all of the needed care, regardless of proof of ability to pay. There could still, of course, be quibbles over what "needed care" encompasses or when the level of trauma reaches extreme, but that is not important for the point I want to make. We might then reach the following assessment of the hospital in this case: [H] The hospital is immoral or unjust because it operates on a proof of ability to pay admission policy that allows for no exceptions in trauma cases such as that suffered by X.

On the account I have been urging [H] is in order as it stands. It needs no reduction to moral assessments about individuals who may have once been or now are the managers or directors of the hospital who wrote, approved, endorsed or enforce the admission policy. [H] is a perfectly appropriate moral judgment of which the hospital *qua* organization itself is the subject. Given what I maintained in Part I, I think we should also say that from our assenting to [H], nothing follows about what C should or should not do, morally speaking, with regard to dealing with X, if, counterfactually, Y had told her that X did not have insurance. Suppose that Y had told C that X was uninsured, would she have acted immorally had she refused to admit X, told the emergency unit to only try to stabilize the patient and send him off on that 150 mile trip to the public hospital that must, by state law, admit the uninsured? Would it have been her moral duty to admit X regardless of the policy, once she was apprised of the severity of his condition? A further complication: suppose that Y took C into his confidence by telling her that the name he had given her for the patient was really his name and that he had done so because he was insured and X was not. Would C have been acting immorally had she joined the conspiracy to get X the needed care from the hospital? I do not think that our answers to any of these questions can be directly inferred from [H], though C's action of either admitting or refusing to admit a trauma patient to the hospital is an institution-constituted act performed within an organizational context.

I have mentioned hospitals, clinics, insurance companies and the like as qualifying as moral agents and so, often, as fair targets *qua* organizations of moral appraisal with respect to their actions. There are other organizations that also concretely embody the institution of medicine and that may also, under certain conditions, be addressed as moral agents in our moral discourse. I have in mind professional organizations, in this country, for example, the American Medical Association. The AMA is organized around a clearly delineated decision structure and it is, through that structure, able to take actions *vis a vis* its membership, address the public at large, carry on political activities, create codes of conduct, render judicial opinions (by its Judicial Council), hold parliamentary sessions (its House of Delegates), etc. The AMA's *Principles of Medical Ethics and Current Opinions of the Judicial Council* (1984) is filled with statements about the actions of the organization *qua* organization. With regard to certain issues it is "unalterably opposed," on others it "encourages," and on still others it "calls upon its members," etc. It threatens penalties for non-compliance, including expulsion.

Consider the following from the *Principles and Current Opinions*: "2.16 Unnecessary Services. It is unethical for a physician to provide or prescribe unnecessary services or unnecessary ancillary facilities." We should probably agree that 2.16 is a morally sound organizational opinion and that its presence in the *Principles and Current Opinions* speaks well for the organization. We might be willing to maintain that the AMA, at least with respect to its code of ethics, is a fair, just, and morally worthy organization and that it is morally important that it exist and police the professional ethics of its members. Any member of the AMA, we could agree, would be morally suspect were he or she not to endorse 2.16 as an important opinion of the organization, though it is conceivable that many members of the AMA, perhaps even a majority, do not endorse 2.16 or believe that it is of any moral import. Nonetheless, it is codified in the *Principles and Current Opinions* and remains organizational policy. The AMA's adoption of 2.16 and member physician P's general agreement that 2.16 is a morally appropriate position for the organization to hold, does not settle, in the sense that a course of institution-constituted action can be unqualifiedly inferred from 2.16, for P whether in a certain set of circumstances P morally ought to perform, for example, an appendectomy on a patient's perfectly health appendix. Those circumstances might be similar to the one's dramatized some years ago in an especially moving episode of the television series *M*A*S*H*. Perhaps by performing the unnecessary operation on a particular patient, P could prevent the potential loss of the lives of a number of other people.

Consider the following argument:

[B] (1) The AMA is a morally just and fair organization in part because it established and enforces a just and fair code of ethics for its members.

(2) The AMA code of ethics includes 2.16, which forbids its members to perform unnecessary medical services.

(3)   Performing an appendectomy on a patient's perfectly health appendix is such an unnecessary service.

(4)   P is a member of the AMA.

Therefore: P should not perform an appendectomy on his patient's perfectly health appendix.

But it is not at all clear that P morally ought not to have performed this unnecessary service on his patient in the case in question: where doing so prevents the potential loss of many lives. (An admittedly utilitarian sounding argument, not here to be discussed.) Further, then suppose that we decide, perhaps following our moral intuitions, that P, in the circumstances, did the right thing by violating 2.16, performing an unnecessary appendectomy. It does not follow that 2.16 ought to be stricken from the *Principles and Current Opinions* of the AMA. Specific cases make bad policies; the AMA's Judicial Council should likely tell us. And the Council would be right. As noted earlier, P might very well endorse 2.16 within the code and maintain that the organization would be morally deficient were it to leave out a statement of the sort expressed in 2.16 from its code of ethics for its members. Suppose P were also a member of the AMA's Judicial Council and he, in fact, voted for the inclusion of 2.16 in the code for the reason stated above: if the organization is to be just and fair it must adopt a statement like 2.16. If P had done so, his vote for its inclusion in the code was about the organization. It was neither a judgment applying to himself, nor was it what Singer calls a "general moral statement" (Singer, 1993, p. 242) of the sort "No physician ought ever to perform an unnecessary medical service." From the general moral statement it does follow that P, being a physician, ought not to perform an unnecessary moral service under any conditions. The confusion of the general moral statement with the judgment about the organization is an example of the "fallacy of misplaced concreteness" (Singer, 1993, p. 242).

The point I think this illustrates is that from the judgment that an organization that is a concrete embodiment of a morally justified institution is just or fair or moral or good, or the like because it has adopted certain policies and/or procedures, nothing directly follows with respect to whether or not an individual member in a specific set of circumstances morally ought or ought not perform certain institution-constituted actions that are either prohibited, permitted, or required by the organization. And, there is a further important implication. Suppose that we agree that P did the right thing in the circumstances by violating 2.16. It does not follow that there is something immoral or unjust or unfair or bad about the AMA because 2.16 is one of its policies as codified in its *Principles and Current Opinions*.

A further complication: Would the AMA be acting unjustly, unfairly, immorally, were it to expel P from its membership for violating 2.16? That is a real puzzler and I am inclined to say that it would not be. My reasons for reaching that opinion are hinged to the need for what we have judged to be a just and moral organization to maintain its integrity as such. In fact, it seems to me imaginable that the AMA Judicial Council might argue that even though P may have done the right thing in the circumstances by performing that institution-constituted action, an appendectomy, he must be expelled for the violation of its code from the organization, and that would further make my point about the gap between moral assessments about organizations and those about the individuals who are members of those organizations.

4. THE MORAL ASSESSMENT OF DIFFERENT ELEMENTS OF THE MORAL WORLD

The picture, albeit still quite primitive, of the moral world that I have been trying to sketch takes as its foundation that there are unbridgeable inference gaps with respect to moral judgments or assessments between institutions and the institution-constituted actions of their concrete embodiments (organizations), between institutions and the institution-constituted actions of individuals, and between the institution-constituted actions of organizations and those of individuals who are members of those organizations. There are also, though I am not concerned with them here, moral assessments of individual actions that are not institution-constituted, and there probably are organization actions that can be the subject of moral assessment that are not institution-constituted either.

Because they are intentional agents, even if organizations are somewhat limited with respect to the scope of their actions, it makes sense to morally assess and hold morally responsible individuals and organizations in terms of both their actions and their characters. Institutions do not act and if we were to talk about the "character" of an institution, we would not be using the term in the same sense that we do to talk about individuals and organizations. Actually, I am not at all sure that we can talk sensibly about the character of an institution in any meaningful way. I suppose there is a sense of "character" that is understood to mean something like "nature" and that we could well ask "What is the nature of a particular institution?" "What is the nature of slavery?" for example, could be answered by saying that it is an institution constituted by a system of rules that yield a certain set of rights, duties, etc. Or we might say that it is an inherently immoral institution because it is constituted by a certain system of rules that yield a set of rights, duties, etc, that violate basic moral standards of human dignity, respect, and worth. In any case, that seems to be a somewhat different matter than assessing Mary's character as racist or as caring, based on our observations of her actions.

Institutions are basic to a social system and the moral assessments of institutions, as in Rawls' theory, are generally given in terms of justness and fairness. Institutions are typically judged as just or unjust, fair or unfair, with respect to their fundamental, though abstract, conceptual structures. To say that the institution of slavery is inherently immoral is to say that it is fundamentally flawed because it is unjust and/or unfair and to found that judgment on defensible principles, typically, regarding the morally appropriate distribution of the goods based in the institution or the society as a whole, or with regard to basic moral principles regarding human dignity, worth, respect, etc. Judging that the institution of medicine is unjust or unfair requires citation to comparable principles. But as the institution of medicine is not inherently unjust or unfair, quite the opposite, the negative assessment would be more in the form of a hypothetical of the sort, [M] "The institution of medicine would be just, or more fully just, if its concrete embodiments, the organizations through which it functions in the lives of the people in the society, were structured to deliver the health care of which they are capable in a just and fair way and they actually did so." In effect, moral judgments about institutions that are not inherently immoral will generally include hypotheticals that relate to the organizations that deliver the goods of the institution. That being the case, we may be tempted to think that no inference gap exists between [M] and a negative or positive judgment regarding a particular hospital and the way it does or does not provide services to, for example, indigent patients. That would, however, be a mistake. From the fact that the institution of medicine would be just or fair (or more so), if the organizations that concretely embody it acted in a certain way nothing directly follows about what policies and procedures a particular hospital ought or ought not to adopt or that that hospital should or should not, in any particular case, admit or refuse admission to an indigent. If we should decide that our institution of medicine is morally flawed, what we are deciding is that our conception of the institution fails to meet the standards of our principles of justice and fairness. We are saying that we need, to use a popular phase, to "rethink it," to reconceive it, until our conception does satisfy those principles. But there is nothing morally or logically incoherent about a hospital, for example, through its board of directors, condemning our institution of medicine as unjust, and at the same time endorsing and enforcing a policy of turning away indigent patients from its doors. We should expect, of course, that once we have reconceived an institution to satisfy our moral principles, we would make efforts to reform the organizations that embody that institution. But organizations operate in the practical sphere, in the real world, not in a world of abstractions, and the gulf, from a moral and a logical point of view, between those worlds is sizable.

Another related point: Unlike organizations and individuals, institutions are not intentional agents. They are abstractions, our abstractions. They do not, cannot, act and they cannot be held morally responsible. They are, to paraphrase Singer and Rawls, relatively permanent systems of social relations organized around a social need or value that are regarded in a society as a way of meeting that need or realizing that value. They involve systems of rules that are built on shared beliefs, values, and purposes within a society. They are not simply customary ways of doing things because customs and habits may not be rule-based.[4] Individuals, collectively, in effect, hold the institutions of their society in their shared world view or, as I would prefer, their shared collection of cares (French and Haney, 2002), and have the capacity to modify most institutions and bring them closer to the principles of justice and fairness by which they are morally measured and to then restructure the organizations that concretely embody them. The reason that I qualified the previous statement is that I do not think they can modify certain institutions, though they can abolish them. I have in mind, slavery, of course, but also promising.

Organizations can act and included in the panoply of their possible actions, as noted above, is planning for organizational change, even the termination of the organization's existence or its merger with another organization. Other organizations and individuals can also effect change in an organization and the same holds true for individuals. The primary point I want to make, however, is that the moral assessment of organizations and of individuals is a conceptually different matter than the moral assessment of institutions. It makes sense to talk of the ethics of institutions and the ethics of organizations and the ethics of individuals without assuming that they can be reduced to a single set of moral principles. Rawls seems to have understood this point. And even if one were convinced that there is a single moral principle that applies to all elements of the moral world that I have sketched, perhaps the principle of utility or the Categorical Imperative, the inference gap between the various elements still would exist. In accord with Rawls' view in *A Theory of Justice*, the principle(s) apply to different subjects and "must be discussed separately" (Rawls, 1971, p. 55).

Briefly, I want to conclude by noting that there is actually a fourth level or element in the moral world that I have not mentioned. That is the social system, comprised of institutions, taken as a whole. Rawls writes: "The primary subject of the principles of social justice is the basic structure of society, the arrangement of major social institutions into one scheme of cooperation" (Rawls, 1971, p. 54). Rawls notes that from the justice or injustice of the social system as a whole, one cannot infer anything about the justice or injustice of any of its institutions or all of its institutions. A social system is an arrangement of institutions according to some rules. Some or all of the rules of arrangement for any given social system may be unjust while the institutions within the system are just. An institution within a social system may be unjust, but the social system containing it may not be unjust, but I suspect that can only be the case when the institution is not a dominant one in the system. Where it is dominant, as was slavery in the antebellum American South, I do not see how it could be the case. Rawls suggests that in some social systems one institution that is unjust might be compensated by another that is just so that "the whole is less unjust than it would be if it contained but one of the unjust parts" (Rawls, 1971, p. 57). It is also conceivable that none of the institutions in a social system are unjust taken independently of each other, but that the system, the arrangement of them into a single system, is unjust taken as a combination. From a moral assessment of the social system and/or its rules of arrangement, therefore, we cannot infer a similar moral assessment about its institutions, nor about the organizations that are the concrete embodiments of those institutions.

There are more elements in the moral world than are dreamt of in an individualist's philosophy.

Arizona State University
Tempe, Arizona, USA

NOTES

1. Apparently, Abraham Lincoln shared this view with Jefferson.
2. See French (2001a) and Kekes (1990) for discussions of "unchosen evil."
3. See, for example, Jackall (1988).
4. I toyed with associating them to what MacIntyre calls "practices," but found his account too broad for my purposes. See MacIntyre (1984, pp. 187-193).

BIBLIOGRAPHY

American Medical Association (1984). *Principles of Medical Ethics and Current Opinions of the Judicial Council*. Chicago: American Medical Association.
Anscombe, E. (1963). *Intention*. Ithaca, NY: Cornell University Press.
Audi, R. (1973). 'Intending,' *Journal of Philosophy, 70*, 387-403.
Austin, J.L. (1970). *Philosophical Papers*. Oxford: Oxford University Press.
Bratman, M. (1987). *Intention, Plans, and Practical Reason*. Cambridge, MA: Harvard University Press.
Davidson, D. (1980). *Essays on Actions and Events*. Oxford: Clarendon Press.
French, P. (1979). 'The corporation as a moral person,' *American Philosophical Quarterly*, 16(3) 207-215.
French, P. (1984). *Collective and Corporate Responsibility*. New York: Columbia University Press.
French, P. (1995). *Corporate Ethics*. Fort Worth, TX: Harcourt Brace Jovanovich.
French, P. (2001a). *The Virtues of Vengeance*. Lawrence, KS: University Press of Kansas.
French, P. (2001b). 'Unchosen evil and moral responsibility,' In Jokic A. (ed.), *War Crimes and Collective Wrongdoings* (pp. 29-47). Oxford: Blackwell Publishing.
French P. & M. Harvey (2002). 'Changes in latitudes, changes in attitudes,' *Contemporary Philosophy, 23* (5&6), 17-28.
Gilbert, M. (1989). *On Social Facts*. London: Routledge.
Gilbert, M. (2000). *Sociality and Responsibility*. Lanham, MD: Rowman & Littlefield.
Goldman, A. (1970). *A Theory of Human Action*. Englewood Cliffs, NJ: Prentice Hall.
Jackall, R. (1988). *Moral Mazes*. Oxford: Oxford University Press.
Jefferson, T. (1950). *The Papers of Thomas Jefferson, 1, 1760-1776*. Boyd J.P. (ed.). Princeton: Princeton University Press.
Jefferson, T. (1954). *Notes on the State of Virginia*. Peden, W. (ed.). Chapel Hill: University of North Carolina Press.
Kekes, J. (1990). *Facing Evil*. Princeton: Princeton University Press.
MacIntyre, A. (1984). *After Virtue, 2nd ed*. Notre Dame, IN: University of Notre Dame Press.
Rawls, J. (1971). *A Theory of Justice*. Cambridge, MA: Harvard University Press.
Singer, M. (1993). 'Institutional ethics,' *Ethics, supplement 35*, 223-245.

KEVIN WM. WILDES, S.J.

INSTITUTIONAL INTEGRITY IN HEALTH CARE: TONY SOPRANO AND FAMILY VALUES

Tony Soprano is someone who worries about tradition and the values embodied in a tradition. He worries about the traditions of his ancestors, his family, and his business enterprises. And he worries a great deal about values like honor and integrity. He often laments that you can't count on people to live out traditional values or keep their word. And, he does his best to instill these values in the younger generation around him.

Tony Soprano is not alone in his worries. He is typical of many contemporary Americans. These concerns reflect a curious, though explainable, turn of events in modern society. In an era when many secular societies are morally permissive, compared with traditional moral communities, and in an era that seems to have few standards of common morality, there are repeated calls for 'integrity' in politics, business, church communities, and health care. These calls for integrity, found in almost every area of contemporary life, are hard to argue with. Who, after all, can be against integrity? Who would say they wanted to lead lives of dis-integration rather than integration?

But, these appeals for integrity, I will argue, are something like a Tower of Babel. People use the same word but the content they give it can be radically different. These calls for integrity are another sign of the fractured nature of the moral life in contemporary secular societies. This essay will argue that integrity is a virtue that must be understood within the context of a moral vision and way of life. In contemporary secular societies there are many different moral visions. In morally pluralistic societies there will be a variety of ways in which integrity is specified. One appeal to integrity may have a different content than another. So in responding to calls for integrity there must be caution as people often have different models and specifications. Such calls run the risk of being empty and turning the moral concerns they represent into slogans or sound bites.

This essay will focus on health care and address issues of institutional integrity in health care. Integrity in health care institutions represents a two fold challenge. First there are the problems of content for the notion of integrity. What is the moral vision underlying the virtue of integrity? Second, there are other difficulties articulating a notion of *institutional* integrity.

Ana Smith Iltis (ed.), Institutional Integrity in Health Care, 29-39.
© *2003 Kluwer Academic Publishers. Printed in the Netherlands.*

Health care provides an important context in which to examine the idea of moral integrity in contemporary secular, morally pluralistic societies. The medical encounter is freighted with moral decisions and meanings. Medicine intervenes and often transforms important moments in human life. The moments of birth, suffering, and death are understood not only medically but morally. These choices and decisions are understood according to people's moral values and not just the presence of certain medical indicators. The moral dimensions of health care and medicine are heightened by the many new possible interventions that medicine can offer a patient. Many key moments of human life, from birth to death, can be transformed by medical intervention. This also means that medical interventions will have moral significance.

The moral dimensions of medicine are not something new to medical practice. They have not been added with the advance of new medical knowledge and technology (Carrick, 2001). These technological advancements have, however, certainly complicated medical decision making. Advances in science and medical technology have created new possibilities that challenge traditional moralities. They have been an impetus to the development of bioethics as a field. This essay will argue, however, that the field of bioethics needs to widen its views to understand many moral dilemmas in health care.

For many reasons the delivery of health care has become more institutionalized and more bureaucratic. These institutional structures often reflect the moral pluralism of modern, secular societies. The shift from individual practitioners to institutional agents is one that is rooted in development of scientific basis of medical knowledge. In this shift the language and focus of ethical analysis has lagged behind. Much of our ethical analysis and language focuses on individuals: individual acts, individual persons, or individual cases. This essay is not arguing against these important focii. But, the essay will assume that this individual focus is an incomplete picture and leads to an incomplete analysis of ethical issues in health care. Individual persons act within a context in health care and the context of health care decisions are shaped by the institution in which health care is delivered. In order to discuss integrity in health care there needs to be an account of institutional integrity.

The introduction of bureaucratic players creates real challenges for doing moral analysis in health care. Because of the power of institutional arrangements in health care, individuals – patients, physicians, nurses, other health care professionals – find the context of decision making and choices limited by the decisions of others. However, within the institutional context it is often difficult to locate moral responsibility in these contexts. A consequence of bureaucratic medicine is that it becomes more and more difficult to speak meaningfully of moral responsibility.

This essay will argue that it is possible to speak of moral responsibility in a bureaucratic environment. To do so, meaningfully, one must understand the context of secular healthcare ethics, concepts of toleration, and the relationship of integrity to mission and identity.

The topic of bureaucratic ethics has received attention in the bioethics literature (Darragh). The rubric of "organizational ethics" has been used to identify and bring together many of these concerns.

It is hard to imagine the shape of modern, scientific medicine to be anything other than bureaucratic. With the role of institutional actors become much more central to the practice of medicine, the role of such actors will become more important for ethics. A focus of ethical questions on individuals – physicians, patients, other providers, etc. – is inadequate because it is incomplete. The actions of individuals, and their moral responsibility, is limited to the context of decision making. To develop a more complete account of bioethics that account needs to include institutional players, their moral responsibility This means that moral concepts and language usually used for individuals must be used analogously and adapted. The challenge for the language of integrity is how to speak of integrity in a morally pluralistic society and how to apply it to institutions and health care organizations.

1. POST-MODERN MORALITY

To understand both the difficulties and importance of institutional integrity one needs to understand the broad contours of contemporary secular ethics and bioethics. There are any number of phrases that are often deployed in contemporary bioethics: post modernism, moral pluralism, secularism (Lyotard, 1984; Engelhardt, 1991). These phrases are important for understanding the context of, and possibilities for, secular bioethics and the challenges of institutional integrity. These terms need to be carefully articulated and used lest they become slogans. If they are developed carefully these concepts can help one understand the context of particular issues in bioethics; especially since so many have to do with public policy.

Post-modernism is a term used in many different fields. It is also a term that has taken on a life of its own far wider than ethics. It is often used to indicate some view of relativism. For this essay it will be confined to ethical inquiry and what it might mean there. To understand post-modernism it is helpful to describe the term "modernity" in relationship to ethics. In the modern age ethics has been part of the philosophical enterprise in the western world. While moral questions and ethics were certainly part of philosophy in the ancient world, in the Middle Ages ethical issues became more associated with theology. In modernity, a post-Reformation age, philosophy became a discipline distinct from theology. As philosophy became more and more modeled on science and scientific reasoning. Philosophical ethics became an appeal to try to find, by appeal to reason alone, a framework for ethics. Indeed Bernard Williams argues that "the moral," as a distinct realm of inquiry, is the invention of modernity (Williams, 1973). The appeal to reason was well known in the ancient world. But in the ancient world moral reason was situated within a culture and an understanding of a way of life (MacIntyre, 1986). However, during the modern age in the west there has been a sharper distinction between ethics, faith, ways of living, and moral reason. Philosophy sought to be a way to think about ethics independent of faith, or a particular community.

This development has led to appeals to reason, outside any particular context or culture, in ethics. Or, it has lead to an appeal to some common faculty of intuition. The appeal to reason or intuition, at first, seems plausible. This appeal to reason was the hope of the Stoics and the Christian natural law tradition. But, in an age that celebrates cultural diversity, it has become more and more clear just how context dependent moral reason and intuitions are. To appeal to reason or intuitions outside a context, or without appealing to a context, is to invite a variety of specifications of moral terms with no way to control them. Each appeal will be embedded in different contexts. Furthermore, each appeal may also have different methods of reason at work.

Since different models will have different content it follows that a post-modern turn in ethics will lead to a morally pluralistic world. A world in which there is not only a pluralism of different judgments about what is morally acceptable but, there will also be a pluralism in how one thinks about both the content and the method of morality (Wildes, 2000). Each method embodies different accounts of moral reason. There are then three levels of moral pluralism. There is moral pluralism at the level of particular judgment. There is moral pluralism in terms of the content of a theory or method. Finally, there is moral pluralism in terms of the different methodologies.

Admitting such pluralism does not lead, necessarily, to relativism. Belief in moral absolutes does not preclude skepticism about the formulation of these absolutes. Here one can appeal to the doctrine of fallibilism, as articulated by Pierce, which holds that absolutes do exist but that epistemic certainty about their formulation is not possible for finite minds (Pierce, 1955). Furthermore, one can argue that even in the midst of such uncertainty there is a minimal framework that shapes the boundaries of any moral position or tradition. Moral discourse is centered on persons as free, consenting moral agents. Persons are the creators of morality. Recognizing that a secular framework does not yield a thick, robust, common morality it does provide a moral framework in which men and women, persons, can cooperate and recognize the moral boundaries of power. Understanding the post modern context of contemporary secular societies helps explain why there are such variations in accounts of integrity.

2. INTEGRITY AND POST-MODERNITY

In view of the post-modern context of ethics, one will approach calls for integrity, in a secular context, with care. Calls for integrity – or the assertion of integrity – seem to be associated with scandals and a lack of integrity where it had been assumed. In the wake of scandals, integrity has been called for in public life (e.g., the 2000 USA presidential elections), in corporate life, in ecclesial life, and in academic life. The public tends to focus on the personalities involved in a scandal (e.g., president, CEO, bishop), and the remedy is thought to be putting the right person in the right position. But, the modern age has been conflicted about the relationship of a person's character and job performance. Machiavelli was perhaps the first to make the distinction and argue that the two need not go together.

However, scandal often goes far beyond an individual and affects a social institution. Often the calls for integrity can be correlated with concerns about trust and the trustworthiness of professions and institutions. One of the claims of the traditional professions has been that they were trustworthy in respect to professional life. Thus, the professions argued for and earned a great deal of autonomy from oversight by society. However, contemporary law, medicine, church, politics, and education have each experienced events that have led the public to mistrust them. There is also mistrust in the oversight mechanisms for these different professions and institutions. To better understand how we might understand integrity in social institutions, we need to understand the concept itself.

One needs to begin with the obvious: defining the term integrity. When we think of integrity we usually conceive it as maintaining a set of moral principles in one's life. One can expand on this conception by thinking of integrity in relationship to the different values (moral and non-moral) and projects that people pursue in their lives. Integrity is more than being consistent; it reflects a ranking of the goals, values, and projects that a person may wish to pursue (Williams, 1973, pp 98-99; Brody & Engelhardt, 1987, pp. 26-28; Brody, 1988, pp. 36-37). Consistency is not hard to do. Fidelity and integrity are much more difficult to achieve.

There are three important observations that need to be made about integrity at this juncture. The first observation is that integrity is more than narrowly following a set of rules or principles. Integrity involves a ranking of values, both moral and non-moral, and projects into the shaping of a life. The second observation is that even with such rankings, one will not be able to use such rankings in a "cook book" fashion to figure out what one should or should not do. Integrity, in this richer sense, relies on judgment and is a matter of discernment. Judgment is not a matter of knowing what is the correct answer. Rather it is a matter of deciding, from a number of possible courses of action, what one takes to be the best course of action. In most cases of judgment there is no single correct answer. The third point about integrity that needs to be explicated is that integrity needs a content. Integrity relies on the values and projects of a person's life to give it substance and content. People often invoke the term integrity as if it had a content of its own. But, as Aristotle argued, integrity is a virtue that helps to order the other virtues. To know the content of integrity, one needs the content of the virtues it is ordering.

In many contemporary discussions of integrity, or calls for integrity, it is often forgotten that integrity is a second order virtue that needs content. It needs the first order virtues, such as honesty and courage, to give it content. This sense of integrity relies on the goals, values, projects a person hopes to pursue. This sense of integrity speaks more of how people pursue their goals, values, and projects. In a certain sense, integrity does not tell someone what the goals should be. What it may do is help someone understand why two goals may be inconsistent with one another. But it does so as a second order virtue.

If one understands integrity as a second order virtue there may be many people who may have integrity but whom one does not admire. This is why this essay began by recounting the worries of Tony Soprano. The central figure of the television series *The Sopranos* is a man of integrity. He is committed to certain values of family, honor, and loyalty. He is tormented, in part, by the erosion of these traditional values in his family. One can see him as a man of integrity while disapproving of the basic projects to which he is committed.[1]

So it is that one must be cautious when responding calls for integrity. Knowing that an institution, or a person, claims integrity does not, by itself, say a great deal. We need to know more. What are the goals, values, projects to which a person or institution is committed? What is the content/substance of the integrity that it possesses? What are the values to which it is committed? To use the term without understanding the vision to which it is related is not helpful. In some circumstances to appeal to integrity, without a content, may be harmful. Different audiences will understand the term in differing ways and reach various conclusions. The use of integrity is similar to another term that is very popular in contemporary writing on education: Excellence. Universities and colleges often herald a commitment to excellence and pride themselves on such a commitment. But one needs to ask excellence about what or excellent in what way? Is there excellence in teaching? Research? Both? What is the vision of education and the educated person that underlies the claims of *excellence*? (Readings, 1996). Like integrity, excellence is a second order characteristic that relies on an underlying content and vision. In the midst of calls for excellence one needs to ask excellence about what.

It may be that claims abut integrity are particularly troublesome in secular, culturally diverse societies. In such societies there are likely to be different understandings of the type of life one ought to pursue. And with these differing views of life, there will be differing understandings of the substance of integrity.

3. INTEGRITY AND COMPROMISE

Discussions about moral integrity are often related to dilemmas of compromise and cooperation. These are particularly problematic for institutions. In an imperfect world one will not always be able to live out the moral life as one thinks it should be lived. Furthermore, in a secular, morally pluralistic world people will often have to work with others who do not hold the same moral views.

If one takes seriously this notion of integrity as a second order virtue, one can begin to articulate and understand how compromise is possible and how the limits of compromise may be defined. For compromise to be possible those involved in the process of compromise must be aware of the values, goals, and projects to which they care committed. Furthermore, they are able to rank these commitments. Such a ranking helps them determine what may be compromised and what cannot be compromised if one is to retain one's basic commitments in an integral fashion.

There is then an intimate relationship then between integrity, compromise, toleration, and opposition.

3.1 Integrity and Compromise

How does this concept of integrity allow us to sketch an account of institutional identity? The first point is that the notion of integrity helps us remember that within a scheme of value commitments there is, usually, a hierarchy of values that constitutes the character of moral projects; that is, some commitments are more fundamental than others to a person and a conception of life. For example, while a person may have a strong commitment to a professional life, there may be an even more fundamental commitment to a loved one or family. For most people, there is some type of ordering of the value commitments they hold. As situations of conflict emerge in a person's life, this hierarchy is crucial to determining which resolutions are acceptable without abandoning the fundamental moral projects which constitute one's conception of the moral life. If there is an internal conflict of values which demands a compromise, one will have to know which commitments are most central so as to know what can, and cannot, be given up in effecting a compromise. The second point is that this model of integrity can develop a model of compromise. A hierarchy of moral commitments allows one to know what can be compromised and what cannot be. It allows one to know which commitments are most important and fundamental.

3.2 Toleration, Approval, and Condemnation

There is a scale of possible responses which institutions can make to the situations and the demands they encounter. The scale ranges from approval to condemnation with a number of responses of toleration which lie between the two extremes. Institutions give approval in acts such as public statements and institutional honors. For example, when a university bestows an honorary degree, or a hospital gives an award, it is usually for some work that a person has done or will do for the institution. There is a sense, however, that such honors "endorse," not only the work that has been done, but also *the person*. That person is honored as a model for the institution. In another example one might think of St. Nowhere's having a hospital chaplain available day and night to care for the sacramental needs of any patient or staff member.   The hospital commits its resources to such a service and may highlight it in its literature and advertising material.   The work of the chaplain touches the core of St. Nowhere's hospital's mission. His work is one which has the hospital's endorsement and approval.

At the other end of the scale of responses, there is condemnation and what might become outright opposition. Here an institution not only withholds approval but expresses strong disapproval by putting all of its resources to work to resist the challenge and refuses to cooperate in any way. For example, if a local law required

St. Nowhere's to perform abortions, one could well imagine the hospital utilizing its resources to fight such a requirement in every possible political and legal arena. A religious university, faced with a law that required the indoctrination of its students in atheism, might commit its resources to a legal battle to overturn such a requirement. Failure to alter the situation may lead to an even more extreme response: civil disobedience and forgoing public funding of any type so that the institution would not be forced to cooperate with evil. It is imaginable that an institution, like a citizen, might employ force to protect its values.

'Toleration' in contrast to approval and condemnation has a very different meaning. It is important to spell out an understanding of toleration. It is often misunderstood in contemporary society to mean acceptance or approval. At its root, toleration means to endure. Many of us endure situations of which we do not approve. Indeed, in many ways institutions that participate in the public domain must tolerate situations that are, at best, incongruous, with their institutional identity.

The use of institutional resources is an aspect of institutional life that is best read through the lenses of this scale. Resources, especially monetary ones, come often from a mixture of public and private resources. Such a funding mixture will carry with it designations and regulations that an institution may have to tolerate. This may well be described as toleration but not approval. In allowing a group to use its resources is not necessarily itself approving of what the group supports. Rather, resource allocation needs to be understood as a very subtle and nuanced form of institutional language; a language that has more voices than simply approval or condemnation.

If one is to understand a notion of institutional integrity then it is important to bring together the ideas of integrity, compromise, and toleration. For health care institutions, these ideas will be crucial for any understanding of institutional integrity.

4. STRUCTURES FOR INSTITUTIONAL INTEGRITY[2]

Health care institutions then find themselves in a situation where there are involved in the delivery of services that are often viewed through moral lenses. They are also in a situation where many of the people who come to use the services, many of those people involved in the delivery of services, may have very different moral views about what is and is not appropriate. In order to navigate these difficult waters, and to make claims about integrity, institution must look at several different aspects of its life. In view of the web that exists between identity, toleration, approval, and resistance, one can better understand how institutions can understand themselves and act in the complex world of contemporary health care in secular societies.

The challenge for health care is how to institutionalize the virtue of integrity. When we speak of integrity on the level of individual persons there are certain elements of a person's life that become central to living a life of integrity. We frequently speak of character, values, and conscience. When one moves the language of integrity to institutions there are other elements that become crucial if the institution is serious about maintaining integrity.

One of the most central and important elements is that of *mission*. If integrity is a second order virtue, then to understand and assess integrity one must be aware of what it is the individual is committed to. There is a difference between the Pope and Tony Soprano even though they may both be individuals of integrity. So too with health care institutions. If an institution is to speak of integrity there needs to be a clear sense of the mission of the institution and what it is about. One can only judge the integrity of an institution if one knows what the institution is committed to and what it hopes to achieve. A mission need not be expressed solely in moral terms. But, if there are moral commitments that are important to the institution, they need to be expressed in the mission.

The statement of mission not only communicates with an external audience, but it also serves the internal life of the institution as well. It is a way to help guide *strategic planning* and create and gage an *internal culture* that is consistent with the mission. Strategic planning is essential in maintaining the integrity of an institution. For one thing, it helps to concretize the mission of the institution. Strategic planning is a way to establish goals and steps toward the realization of the mission. In a strategic planning process the mission becomes a way to both guide and evaluate the planning.

Another part of the planning process is the creation of a *budget*. A budget is essentially a planning document that expresses how an institution wants to use its resources. For example, if an institutions says, in its mission, that it is committed to the care of the poor yet provides no line item in the budget for uncompensated care, there is a serious gap between the mission and the strategic planning. There ought to be an integration between the mission of an institution, its strategic plan, and its budget. The budget of an institution ought to express the values and mission of an institution.

The mission of an institution ought to shape the internal culture of the institution as well as how it deals with the external world (customers, patients, etc). There are two good reasons to work on the internal culture. One is that it creates a deeper level of integration. Another is that it creates a deeper level of buy in within the internal culture to the mission as a whole. A key question in creating a internal culture that supports integrity of mission is the question of who has *voice* within an institution. An institution needs to create processes that give members voice in the planning processes and voice in the *assessment* of planning and how well goals are being carried out.

In all the discussion of integrity one also needs to examine who has responsibility for the integrity of the institution. As an institution articulates its mission and goals, its way of proceeding, there needs to be some one(s) who bear responsibility qua an officer of the institution for decisions that are made and the assessment of the processes that can help to insure integrity.

It becomes possible then to think about integrity and identity in the complex world of health care in secular societies. The place of mission is central.

Fashioning, revising, deepening, renewing this self-understanding is crucial to any account of integrity. Only in light of mission can the other tools be understood (e.g., planning, budgeting, culture) and implemented. Also, only in light of mission can a health care organization respond to other institutions, relationships, and their values. Only in this light can compromise and cooperation be understood and implemented.

5. CONCLUSIONS

It is understandable why people would desire integrity in their own lives, the lives of others, and society. In a morally pluralistic world, integrity comes to play a very important role. In a world where people, and institutions, can live and work out of very different moral visions, integrity will be an important element for cooperation. When one cannot assume that a person, or an institution, shares one's own moral values, then the key to successful cooperation will be that people and institutions clearly announce, and live out, their basic moral commitments. If institutions live out a virtue of integrity, then that will enable those who work for them (physicians, nurses), those who use their services (patients, health care plans), and those who partner with them (other health care institutions and organizations) to do so with informed consent. One might think of institutional integrity along the lines of the practice of informed consent – a practice that has become essential for navigating moral differences in contemporary health care. Institutional integrity will be the virtue that underlies an institutional practice of informed consent. In this way employees, patients, and partners will be able to cooperate with an institution and the values it holds.

Georgetown University
Washington, D.C., USA

NOTES

[1] The points I develop here are the development of earlier work in my writing and lecturing on this topic. My thoughts have continued in the same line but I have become more aware of the depth of the issues (see Wildes, 1991). I have even moved my cultural allusions from The Godfather to Tony Soprano.
[2] These points also grow out of earlier work that I have done (see Wildes, 1997).

REFERENCES

Brody, B. (1988). *Life and Death Decision Making*. New York: Oxford University Press.
Brody, B. & Engelhardt, H.T., Jr. (1987). *Bioethics: Readings and Cases*. Englewood Cliffs, NJ: Prentice Hall.
Carrick, P. (2001). *Medical Ethics in the Ancient World*. Washington, DC: Georgetown University Press.
Darragh, M. (1997) 'Ethical issues in managed care.' *Kennedy Institute of Ethics Journal* 7:421-426.
Engelhardt, H.T. (1991). *Bioethics and Secular Humanism: The Search for Common Morality*, Philadelphia: Trinity International Press.

Lyotard, J.F. (1984) *The Postmodern Condition*. Manchester: Manchester University Press.

Readings, B. (1996). *The University In Ruins*, Cambridge, MA: Harvard University Press.

MacIntyre, A. (1986) *After Virtue*. Notre Dame: University of Notre Dame Press.

Peirce, C. S. (1955). 'The Scientific Attitude and Fallibilism.' In: Buchler J. (ed.), *Philosophical Writings of Peirce* (pp. 42-59). New York: Dover Books.

Williams, B. (1973) 'A critique of utilitarianism.' In: Smart J.J.C. & Williams B. (eds.), *Utilitarianism: For and Against* (pp. 76-150). London: Cambridge University Press.

Wildes, K. (1991). 'Institutional Integrity: Approval, Toleration and Holy War Or "Always True to You In My Fashion".' *The Journal of Medicine and Philosophy 16*: 211-220.

Wildes, K. (1997). 'Institutional Identity, Integrity, and Conscience.' *Kennedy Institute of Ethics Journal 4*: 413-419.

Wildes, K. (2000) *Moral Acquaintances: Methodology in Bioethics*. Notre Dame: University of Notre Dame Press.

RONALD C. ARNETT and
JANIE M. HARDEN FRITZ

SUSTAINING INSTITUTIONAL ETHOS AND INTEGRITY: MANAGEMENT IN A POSTMODERN MOMENT

1. INTRODUCTION

A local hospital runs a campaign for being a place of "caring and healing". In conversation with a consultant for that hospital, one hears contradictory insights. Employees feel underpaid in a hospital strapped for money, unable to supply the latest equipment. The advertising misrepresents the "real" story that guides the hospital. Public information has managerial and employee consequences. Cynicism, fueled by disconnection between public mission and organizational reality, decreases managerial effectiveness and endangers institutional integrity and ethos. Health care institutions need communication that promotes organizational health, lessening the possibility of organizational cynicism.

One of the greatest challenges facing health care institutions is maintaining a story or narrative that identifies the institution's core values (mission) (e.g., Clark, 1972), supported by appropriate communicative social practices that give life to those values (i.e., alignment of mission and structure) (Bart and Tabone, 1998). When employees sense that nothing but profit drives an organization, espoused values of an organization's culture, mission, or story lose their currency and power to offer a "why" to work within a particular organization. For an organization's external audience as well, failure to match word and deed leads to cynicism or mistrust in the organization's institutional integrity and ethos (Bruhn, 2001). Cynicism dwells in the land of mistrusted and unknown "whys".

Cynicism plagues organizations today (Dean, Brandes, and Dharwadkar, 1998; Kanter and Mirvis, 1989), particularly those in the health care industry (Bruhn, 2001; Burke, 2002), where calls for a renewed emphasis on professionalism and ethical behavior are increasing in the face of decreased public trust in health care institutions (Bruhn, 2001). Individualism, characterized by employees not thinking

Ana Smith Iltis (ed.), Institutional Integrity in Health Care, 41-71.
© *2003 Kluwer Academic Publishers. Printed in the Netherlands.*

beyond "me" and taking care of the self, is a consequence of cynicism; if others cannot be trusted to look after "me", I must do so myself. Cynicism and individualism are exacerbated in an era of corporate downsizing; health care institutions suffer decreased employee commitment as trust in organizations declines (Spence Laschinger, Finegan, and Shamian, 2001).

Managers of health care institutions take on the major task of assisting with institutional direction, bearing responsibility for implementing the story of a particular health care institution. An organizational story offers clarity, guidance, and restraint for internal and external audiences, setting parameters for judgment and action (Arnett and Fritz, 2001, p. 183), and directs an organization's understanding of its role in the community (e.g., Green, 1990). An organizational story serves as a "common center" (Buber, 1958, p. 89), pulling people of difference together: Why work for this organization? An organizational story presents a unique institutional identity to external audiences, permitting differentiation of this organization from others: Why come to this institution for health care? A consistent organizational narrative frames life in organizations by guiding communicative social practices that support and reinforce the mission or story, sustaining the organization's ethos and preserving its integrity. An organizational narrative guides through the challenges posed by different historical moments, suggesting lines of action appropriate to meet the need of the historical moment without putting at risk the organization's identity. Narratives offer argumentative parameters for employees and for the organization, providing necessary restraining limits. A story anchored in more than profit assists organizations in fulfilling a role as community citizens, restraining unethical action.

For employees to believe the institutional narrative and to support it with action and commitment, rather than responding with cynicism and a focus on the self, managers must lead the way, with supporting action on the part of those who own and direct organizations (e.g., Fritz, Arnett, and Conkel, 1999). If employees do not support or "buy into" the story or mission with accompanying needed social practices, then the institution's identity and integrity will be compromised (Bisson, 2002). Furthermore, managers must direct employees' focus of attention away from the "me" of individualism and toward the organization's narrative by reclaiming the organization as a public place where a communicative ethic of professional civility guides task-focused behavior (Arnett and Fritz, 2001). As employee behavior supports the organization's identity, organizations preserve institutional integrity and win the opportunity to regain public trust, the lost legacy of professionalism (Sullivan, 1995).

Cynicism and individualism facing institutions are not unique to this historical moment. They have appeared before, rooted in changing philosophies of identity across historical moments in the United States. By examining the implications of these historical changes for organizations and management, we invite consideration of the case for organizations as invitational narratives and managers as storytellers within an ethical communicative framework of 'professional civility', an accompanying public communicative ethic that guides implementation of an organization's narrative, with implications for employees, organizations, and the larger public of which they are a part. Our call to understand organizations as narratives and managers as storytellers springs from our reading of the postmodern moment, a time of contested virtue structures and lack of agreement on larger narratives that offer clarity, guidance, and restraint to individuals and institutions. In postmodernity, no one virtue system/story can guide a people (MacIntyre, 1984). Stories as propellers of direction exist within a milieu of competing rhetorics, each vying for the opportunity to lead. A narrative is a rhetorical construction capable of persuading and guiding a people. A story is a beginning effort to persuade; a narrative carries the weight of the commitment of a people. For purposes of this essay, we will use "story" and "narrative" interchangeably with reference to organizations.

We articulate the need for organizational narratives that give meaning to information and work, calling for attention to something beyond "me" and profit. Narratives offer missing clarity, guidance, and restraint to persons and organizations, but can be sustained only within a background context of professional civility offering guidelines for implementing an organization's narrative. Professional civility is an "implementation" story of organizations as public places within which respect for one's "local home", for professionals with whom one works, for productivity, and for the larger public propel organizational life, offering clarity, guidance, and restraint in the process of telling an organization's story.

We begin by addressing cynicism in today's workplace, tracing its historical rootedness in times of transition and change and its subsequent connection to individualism. A shift of metaphors from place to self took place with the movement from an agrarian economy to an industrial economy, moving the locus of meaning for life and work, with cynicism functioning as a rhetorical signal of transition from one historical moment to the next. We then move to a framework for management based on the concerns of (agrarian) place, (industrial) self, and (postmodern) story, articulating the rhetorical task of management in a narrative organization during a postmodern moment as giving direction through storytelling. We end by describing a public communicative ethic of professional civility, a story to guide implementation of an organization's narrative.

2. THE HISTORICAL MOMENT: A TIME OF CHANGE AND CYNICISM

2.1 The Message of Cynicism

Cynicism is a reaction to distrusted direction. Routine cynicism (Arnett and Arneson, 1999, p. 18) emerges when a given rhetorical direction no longer invites confidence, but engenders rejection, distrust, and perception of lack of veracity. That which is stated no longer evokes trust. Kanter and Mirvis (1989) outline the nature of cynicism in the workplace:

There are three key ingredients in the development of the cynical outlook. One is the formulation of unrealistically high expectations, of oneself and/or of other people, which generalize to expectations of society, institutions, authorities, and the future. A second is the experience of disappointment, in oneself and in others, and consequent feelings of frustration and defeat. Finally there is disillusion, the sense of being let down or of letting oneself down, and, more darkly, the sense of being deceived, betrayed or used by others. Turning their disillusion inward, cynics fear they might be seen as naive or be taken for suckers. (p. 3)

Failed high expectations and lack of trust in direction invite a pervasive cynicism, challenging commitment to organizational mission/story, and discounting a reason for engaging particular communicative social practices. Cynicism can be an appropriate response when a given rhetorical direction is no longer accepted, believed, or supported by communicative social practices enacted in good faith. Cynicism is the public signal that an audience no longer perceives congruence between words and action of a given speaker/organization (e.g., Reiser, 1994).

There are two philosophical keys implied in the above position. First, reflective awareness of a given direction frames the *why* or rationale for a given direction. An accepted sense of *why* propels action. Strategies devoid of *why* will move an idea forward until difficulty causes one to ask, "*Why* am I doing this?" Minimizing cynical discourse requires rhetorical direction that includes a verbalized *why* accompanied by congruent communicative social practices. Second, cynicism is a rhetorical device that originates with the audience; it is a form of audience feedback announcing disparity between speech and action. Examined in this fashion, cynicism is a form of rhetorical currency in the workplace; cynicism has message value. Cynicism is not socially dysfunctional at all times. This essay questions routine cynicism, not historically appropriate cynicism that signals disparity between speech and action, indicating a need for a rhetorical shift of direction in either oral discourse or action.

Although cynicism has an appropriate time and place, two warnings about the phenomenon of cynicism guide this essay. First, cynicism becomes problematic when it assumes the position of unreflective, routine communicative action. Cynicism as a response to every organizational action generates a formulaic response: "There they [those who run the organization] go again." Second, cynicism is not disagreement. Sometimes rhetoric *does* manifest congruence between speech and action, but meets disagreement from the audience. This disagreement invites conflict, not cynicism. Cynicism is awareness of disparity between word and deed. Conflict grows from disagreement with action appropriate for a given historical situation. When cynicism is used to describe genuine conflict, it can diffuse examination of real differences, placing the discourse in a context of "they [these employees] are always cynical." This manner of avoiding conflict is similar to framing a genuine, issues-based conflict within the language of a "personality conflict".

This essay recognizes the socially constructive rhetorical contribution of cynicism that announces incongruity between a speech act and later actions. Sometimes a given employee cannot prosper in a particular institution due to incompatibility between rhetorical direction of a company and the manner or direction in which a person wants to work; "supplementary" person-organization fit (Kristoff, 1996) may be poor. Such a state invites disagreement, not cynicism. Dangerous or routine cynicism equates misalignment of story and communicative social practices with every act of disagreement. Such interpretation is a form of cynical "listening", embracing "psychologism", the attribution of motive to the speech of the other, even when it is not present. The false safety of a routine deconstructive technique, based not on the text of the other's speech, but on one's own hermeneutic of suspicion and attribution, then bypasses recognition of authentic expression of rhetoric that is congruent in speech and action.

Cynicism is conveniently used as a defensive response to change. As eras change, displacement occurs—not necessarily the ground for genuine cynicism, but often the soil for routine, unreflective cynicism.

2.2 Cynicism and Displacement: Shifting Eras

Significant social conflicts accompany movement from one era or "wave" to another (Toffler and Toffler, 1993). Change disrupts taken-for-granted patterns, often calling attention to the previously unnoticed or unreflectively appreciated in daily living. The movement from an agrarian period to an industrial era to a postmodern age has not been without anguish and pain; change is seldom desired when we are content. Movement from one era to another disrupts our well-honed practices, or "habits of the heart" (Bellah et. al., 1985; de Tocqueville, 1966), to which we have become accustomed, requiring us to learn new skills, new ways of associating together. Chaos emerges as normative in such times of transition. Chaos is a period of adjustment, action without a *telos*, a necessary waiting – waiting for new patterns to take hold that might offer a sense of clarity and direction:

> Because massive changes in society cannot occur without conflict, we believe the metaphor of history as "waves" of change is more dynamic and revealing than talk about a transition to "postmodernism." Waves are dynamic. When waves crash in on one another, powerful crosscurrents are unleashed. When waves of history collide, whole civilizations clash. And that sheds light on much that otherwise seems senseless or random in today's world. (Toffler and Toffler, 1993, p. 18)

Change makes more sense from afar than from the moment of endurance. Fundamental change forced upon us that challenges our competencies is rarely met with eager acceptance.

Toffler and Toffler suggest that war often emerges when new waves of change come too quickly. Current power structures react to new waves of change that shift power, no longer attracting the same social rewards in a new age. For instance, an agrarian age required muscle, rewarding physical strength. An industrial age centered upon the machine encouraged knowledge of machines. A postmodern age empowered by information calls for yet another set of life and work skills. The skills of one era do not necessarily correlate with the needs of the next, inviting cynicism as disparity between the needs of the new historical moment and worker skills become increasingly obvious.

The invitation. The invitation for cynicism in a transitional era emerges when the rhetoric one has absorbed about the "good life" and the skills one employs no longer are congruent with the needs of a new historical period. Note that both a (now outdated) view of the "good life" and one's (obsolete) skills are incongruent with the requirements of a new era. Such disparity generates understandable anger. In this case, the "speaker" is the historical moment of a new era and the audience is a person with "old" skills and a dated view of the "good life". The new historical moment announces a new era, but the social practices that permit that era to work well are not yet in place; all that exist are social practices from the previous era. These old social practices lag behind the rhetorical demands of a new historical moment, inviting cynicism from perceived disparity between speech and action.

The rhetorical signal of an organization entering a new historical moment is cynicism. Workers sense that what they understood as the "good life" within a given organization and the skills needed to attain that sense of the "good life" no longer connect with the historical needs of a company. Again, the notion of cynicism can be viewed as constructive – a rhetorical message that a world view is undergoing sizeable change. At its best, a rhetoric of cynicism invites us to reconnect with historicity rather than to wallow in nostalgia, calling us to meet a new historical moment, not to rely on the comfortable and known.

Transitions from one era to another do not happen immediately. However, the transitions from an industrial age to a postmodern age took much less time than the movement from an agrarian to an industrial age. The United States population, situated in small towns and farms as we entered World War II, quickly shifted to the cities. We then moved from an agrarian to an industrial to a postmodern era in less than fifty years. Is it any wonder that many feel pushed by the speed of change?

Hannah Arendt (1958/1989) describes some reasons for our movement from an agrarian age. She examines Descartes's call to radical doubt, suggesting that Descartes's message paved the way for the industrial revolution. An agrarian worldview based on *place,* lack of mobility, and unreflective action composed of well-developed practices established a sense of a predictable pattern. The industrial age required workers' displacement from the farm, resulting in gathering on one site, away from family, friends, and familiar surroundings. No longer was it possible to do all in a familiar place with unreflective practices.

Doubt offers a conceptual tool for questioning a sense of place. Doubt is a baseline ingredient for change and cynicism—what was previously accepted is no longer believable. Doubt made possible our understanding of science and modernity. Descartes was a messenger for change; he offered a rhetorical convention that permitted us to recognize change and pave the way for another era--radical doubt of the taken-for-granted. In ferment of change, cynicism becomes an act of conceptual clarity, permitting one to see the before unseen. The notion of radical doubt was necessary for Marx to call the proletariat to reexamine social structures. Descartes unleashed the industrial revolution and its critiques, permitting human agency to question structure already in place. Doubting opens eyes to disparity of word and action, signaling the arrival of a new world view.

Sennett (1974) examines the move from an agrarian to an industrial era by exploring the shift from an aristocratic culture to a bourgeois culture. He discusses "the displaced," people thrust off farms and into the cities as "strangers":

> In the population of cities, a special sort of stranger played a critical role. He or he was alone, cut off from past associations, come to the city from a significant distance. . . . there seems to be no social order among them. And since they have no form, he [Defoe] expects them to wash away from the city in the same way they have come. . . [yet they keep coming and they stay]. (p. 51)

Migrating to a city in search of work only to find rejection invites cynicism. Perhaps many came to the city with an understanding of community or the "good life" grounded in an agrarian sense of place, only to find their old skills of little assistance in the coming era. Additionally, the new city arrivals were not wanted. Cynicism, as a legitimate and common communicative style appropriate for lives depicted in the novels of Charles Dickens, revealed a dream contrary to what immigrants envisioned. The two largest cities at that time, Paris and London, were growing rapidly from the arrival of more and more "strangers" who found little of a sense of place, increasing the numbers of the disenfranchised.

Displacement as home for cynicism. Displacement is the home of cynicism. The movement from an agrarian sense of place to an industrial era with the focus on self shapes Upton Sinclair's *The Jungle* (1905/1960). The novel centers on Jurgis, uprooted from his Lithuanian "ground", displaced, exploited, and abused in the Chicago meat factories, who embraces the metaphor of *self* as a survival framework for the industrial revolution; the notion of *self* replaces the metaphor of *place* in the new era. After losing his home to penury, his wife to an early death, his freedom to the "punishment" of imprisonment, and his health to unwholesome working conditions, what Jurgis had left was himself. Interestingly, Sinclair introduces Jurgis to union life and socialism as a way to counter the exploitation. Sinclair points to the only way out of the *self* focus, a *philosophical sense of home* (discussed later in detail). One needs to be part of a tradition larger than oneself. Writing years earlier than our current situation, Sinclair suggested what postmodernity now makes clear--our task is to frame lives within stories worth knowing, worthy of our participation. In the words of Walter Fisher (1987), we are *homo narrans* (p. 62).

Is it surprising that cynicism would emerge in moments of fundamental transition from one historical moment to the next? Cynicism is often a legitimate communicative act for the displaced. Cynicism has a social function: it announces movement from one historical moment to another.

The shift of metaphors from *place* to *self* is powerful in *The Jungle* and present in lives moving from the agrarian to the industrial age. A clear description of the focus on *self* in the industrial age offered by Christopher Lasch (1984) provides an understanding of the transition from self as an act of survival to hyper-reflection on the self, leading to narcissism. The movement to narcissism arises when genuine crisis to the self is no longer differentiated from problems of a typical day:

> The trivialization of crisis, while it testifies to a pervasive sense of danger – to a perception that nothing, not even the simplest domestic detail, can be taken for granted--also serves as a survival strategy in its own right. When the grim rhetoric of survivalism invades everyday life, it simultaneously intensifies and relieves the fear of [genuine] disaster. (p. 62)

Lasch's "minimal self" is a fortress, a cynical defense against a world displacing workers from a sense of place into an industry of exploitation in organizations with only the self left as defense. Today, cynicism about "I" driven language and focus on the self stems from the unfulfilled promise of a therapeutic orientation to life that promises potentiality of the self with no sense of limits. Calls for civility (e.g., Carter, 1998; Arnett and Arneson, 1999) and other alternatives to cynicism are emerging in response.

What metaphor guides a postmodern era, when both *place* and *self* belong to different historical moments? To manage today requires understanding what metaphor reflects the communicative social practices of a postmodern era. Metaphors emerge out of our social practices; an emerging era requires rhetorical trail blazing and testing as we articulate metaphors potentially capable of guiding in a postmodern age.

3. MANAGEMENT IN A POSTMODERN MOMENT

Postmodernity is an era of narrative contention and virtue disagreement, a time of "profound cultural changes which have conspired to bring institutions and their authority into question" (Engelhardt, 1990, p. 34), leaving few guidelines or insights on how to manage. When confused, one can examine the communicative social practices that seem to assist this historical moment and reject those social practices no longer appropriate. From the social practices of our era, MacIntyre (1984) and Bellah and colleagues (1985) describe the character types appropriate for modernity and for an agrarian culture. Each agrees that the manager is a major key to understanding modernity.

Management did not begin with modernity and the industrial revolution, but it did become a profession in that era (Kimball, 1992). Management in an agrarian age took care of a place. Often one worked with a company registered under the family name. Management in modernity watched and directed the worker. The metaphor of self played out in the management of the other, the "other self". Management continues in postmodernity, but what communicative metaphor seems most appropriate in such an era? The metaphor of story, not place or self, might offer guidance for management in a postmodern era.

3.1 Management: Offering Direction

Management has moved from managing place, to managing the selves of workers, to managing the story of work in a given organization in a particular historical moment.

Of course, the metaphors of place and self continue to shape a manager's task. What is different is that these two metaphors no longer rest as first principles; story takes on first principle status, offering a framework that fits the needs of an organization at a given point in time. Each era required a manager to offer direction: direction for a place, then for the selves in the organization, and now for the task of work itself within the organization. The next section frames the importance of managers telling a story of "why" one would labor in a given organization and offers the public communicative social practices that support such a story.

By examining the practices of various forms of management, we find that the common rhetorical feature is direction. One communicates direction no matter what era one inhabits or what style of management one employs. In the authoritarian or classical structure of Taylor's scientific management (Taylor, 1911), the manager dictates to the other to proceed in a given direction. In a human relations or human resources model, one facilitates a direction. In each case, there is a common concern with direction and keeping on task. But in each case, some form of direction is provided. In a small business in which the manager is also a laborer, the key is modeling a direction; small business managers model communicative social practices that shape organizational stories. This essay suggests that small business, in its emphasis on "doing" communicative social practices, offers a key for negotiating a postmodern age.

Returning to the threefold eras of agrarian, industrial, and information ages, we imagine a farmer as a manager of sorts. We discover the agrarian age typified by direction by example (*participation/modeling*) – working in fields next to family and known employees. In an industrial age, a manager is more of a *watcher/critic* than a participant. Direction emerges in an agrarian age by participatory action and modeling. Direction in an industrial age requires verbal correction, little participation, and little direct modeling. From an agrarian to an industrial world view we move from rhetorical action as that of the *unreflective participant/modeler* (one cannot think of doing anything other than working alongside the other) to that of the *reflective spectator/critic* (one verbalizes alternative actions for the other), which result in two distinctly different approaches to rhetorical direction. A small business continues to work in a manner akin to that of an agrarian age, calling forth participation from all members, modeling key communicative social practices. The professional manager in an industrial age does not engage a task in the same intimate fashion as the participant/modeler. As we move from an agrarian to an industrial era, there is increasing distance between a manager and the product that, through managerial direction, workers produce. The "Other's face" (Levinas, 1987, p. 113) becomes more distant and labor more remote (Arendt, 1958/1989). Marx's notion of "alienated labor" (1847/1975, p. 77) is more complex than this description allows, but it begins with awareness of increased distance between the person and the labor itself as work is done for another in the form of "surplus labor" (pp. 90-102). The move to a more mechanized and impersonal workplace resulted in the emergence of the "minimal self" (Lasch, 1984) as a legitimate way to enhance the probability of survival. The "minimal self" implies taking care of oneself. Cynicism, as a rhetorical message, suggests that it may be necessary once again to alter rhetorical direction, to seek a new set of guidelines for rhetorical direction that are historically distinct and appropriate. What is lost from an agrarian era that remains as a remnant in small business is participation and modeling of communicative social practices that invite an organically grown story of the life of the organization, its values, and its direction.

We determine appropriate rhetorical direction by observing social practices that work well. A college president watched where people walked before building sidewalks on the footpaths. The footpaths of *participant/modeler* and *spectator/critic* are clear. Since social practices are not yet in place in this new moment, we stray from observation of social practices and suggest that the yet-untrod footpath of a *reflective participant/storyteller* might be the rhetorical key to our coming age, marking the parameters of a small business model.

As we enter this information age, this postmodern age, with no clear rhetorical direction, the answer needs to reveal itself in the pattern of social evolution. We are just entering this new era. Only reflection on social practices of this era after they are commonplace will offer genuine clarity. However, we can find guidance. This essay suggests that within an agrarian and a more recent small business model, we may find assistance for story construction in a cynical age. Participation and modeling may be the only ways to invite a story that people can agree to and with in an age of cynicism that makes an art of a "hermeneutic of suspicion". Our task is not descriptive or prescriptive, but suggestive and invitational, re-examining the metaphors and the guiding communicative social practices of an agrarian age that continues in remnant form in many small businesses that offer life to the metaphors of participation and modeling.

We make a rhetorical turn, outlining communicative social practices within the metaphors of *participant/modeler* and *reflective storyteller*. Unlike an agrarian effort that used a participant/modeler orientation unreflectively, we must do *reflective participant/modeling*, which lays the groundwork for the story, the sense of "why", of a given organization.

3.2 Management: A Rhetorical Task

Movement from an industrial era into an information age takes us from clarity of social practices to the reality of little guidance from yet unframed social practices appropriate to a new historical moment. A new era does not come with social practices in place; they must evolve. Discerning an answer to the question of "How will we discover appropriate communicative social practices in an information age?" becomes a rhetorical or persuasive task. The historical moment functions as a suggestive rhetoric, encouraging us to connect an emerging era to appropriate communicative social practices.

Management understood as a rhetorical task (e.g., Tompkins and Cheney, 1983) requires finding language that connects needed social practices to the historical moment, which includes the audience, the people within an organization.   A management perspective is not culturally neutral, but rhetorical, situated within persuasive bias.   As we engage organizations, we need to be rhetorically sensitive to the presence of multiple eras in the same historical moment.   This is the fate of postmodernity. Clarity requires attention to the particular, not the universal, in such an age.

For example, an agrarian age is both past and co-present in a postmodern age, functioning as both a historical artifact and a live reality within this historical moment. Additionally, an industrial age is both past and co-present. As we entered World War II, many would have said that the United States was in an industrial era, but the percentage of people living in small towns and on farms demonstrated the lasting power of the agrarian era, as did the Mom and Pop store in modified form with an emphasis on participation and modeling of work. Historical eras do not die easily; they enjoy a long walk with eras that follow. A postmodern moment finds both an agrarian and an industrial age co-existing in the present.

As the industrial age gathered momentum, management and labor unions developed rules that institutionalized distance between management and labor. The industrial age brought us the "white collar" worker, just as the agrarian age brought us the "farmer." The farmer managed the ground and the white collar worker began to manage the shop. The task of managing the shop resulted in more distance between management and labor than managing the farm, where getting one's hands dirty alongside the other typified managing the ground.

In large organizations, the face-to-face management of farm life of years ago is less possible with large numbers of employees. Organizations began to eliminate middle management positions in the 1980s and 1990s (Snizek and Kestel, 1999, p. 63); the role of watcher/critic is less valued as we move from this industrial norm. The farmer manager told a story by action, and in a postmodern age we need once again to tell a story, but a story appropriate for a postmodern age.   This essay suggests that the rhetorical task of the manager today is to frame a story that unifies around a publicly proclaimed direction that outlines the *how* (the social practices) consistent with the *why* (the public story of an organization).

3.3. Manager as Storyteller in the Narrative Organization

Storytelling in an information age offers direction without dictating or assuming that one meta-story offers universal guidance.   Modernity brought us the metanarrative, the universal. Postmodernity brings us "humble narratives" (Arnett, 1997a, pp. 44-45) that provide temporal direction, momentary guidance. In an agrarian age, one guides by participation. In an industrial machine age, one guides by increasing distance in a supervisory role. In an information age that rests within an era of postmodernity, guidance emerges from historically appropriate good stories—stories that give meaning to disparate pieces of isolated information. We live in a moment in which information exists without connecting links of meaning, creating a communicative context where storytelling must frame information and our participation together.

Storytelling in an information age offers a philosophical sense of place that moves beyond a physical sense of place and beyond protection of the self to a story that provides a temporal "common center" (Buber, 1958, p. 89). A story gathers action, provides a sense of *why*, and offers rhetorical direction with opportunity for participation and meaningful construction of information. The manager as a storyteller does not focus upon *place* or *self,* but offers a story that gives a sense of meaning to the effort of work, *a philosophical sense of home.* Storytelling is an effort to restore work (participation and information) with meaning beyond daily routine and individual survival.

Storytelling is not new. We find the effort to add meaning to participation and information in documents ranging from the Bible to the Declaration of Independence to *Das Capital.* Narrative theory and the importance of storytelling in the construction of meaning are increasingly recognized in management and organizational studies from both critical and functionalist perspectives (O'Connor, 2002). In business, the buzz about mission statements (Swales and Rogers, 1995) tells us of hope for a way to coordinate information and meaning in a given organization. Mission statements can be attempts to move meaning beyond the routine and individual survival.

Examining the mission of a company from the standpoints of agrarian, industrial, and information ages reveals the changing role of management in the construction of story-laden meaning. In a modeling/participatory agrarian age, there is no need for a mission statement. Family and known persons who work alongside each other hold the company together. As the company becomes more industrialized, the spectator/critic watches over large numbers of persons. In the first instance (agrarian), rubbing shoulders with the other guides, and in the second, fear of the critic directs action. In both cases, it is a person with some face contact who guides. In an information age, as face-to-face contact is lessened, employees need to have a new sense of why for action and a why for the organizing of information. If neither face-to-face modeling nor a focus upon the self assist in organizing a unit, then perhaps it is time to offer a story that reframes participation and information. Management as storytelling (e.g., Boje, 1991) is a pragmatic response to the workplace in an information age.

Storytelling presuppositions. Storytelling in an information age is also tied to another description of our era – postmodernity. Up until this point, the notions of an information age and postmodernity have been used almost identically. Clarification of the two terms is necessary. Postmodernity is better understood as a juncture, not an era. Postmodernity is the transition moment between a dying era and the next. The information age is a functional description of an era replacing the industrial age.

This essay relies upon central metaphors that open our interpretive participation with the question of management in this historical moment. In particular, the terms agrarian, industrial, and information ages guide our analysis of changing demands in management. We assume that an information age rests within postmodernity, an era without a guiding narrative structure. The rationale for this assumption is simply that our capacity for making or acquiring information has run faster than our ability to situate information within meaningful contexts. Postmodernity is a reaction, not to the industrial life of modernity, but to the efficiency of information construction of modernity that did not take seriously enough the necessity of situating information within a meaningful context. This essay offers the notion of story as key for resituating information within a meaningful context and offering a reason, a "why", for work and participation.

Postmodernity calls into question two basic assumptions: first, the supremacy of agency, the ability of the metaphor of self, current in an industrial age, to guide us in an emerging historical moment (Best and Kellner, 1991, p. 5; Rosenau, 1992, pp. 42-44). Second, postmodernity suggests that we cannot agree on a common narrative or virtue structure to guide us (MacIntyre, 1984). Modernity relied upon a common value system based on the myth of progress, outlined well by Lasch (1991). An agrarian world view's tacit agreements about place invited by the farm community and small town based on family, religion and patriotism are also disrupted in a postmodern moment. Postmodernity calls into question both the foundations of an agrarian age and an industrial age emphasis on self.

The notion of an information age is an appropriate companion to postmodernity. We assumed that information guides; it does not. Meaningful information guides--nothing less. With an inability to count on the meaning centers of *place* and *self*, postmodernity calls for another metaphorical center of meaning – stories. Stories frame information, offering a philosophical sense of place that organizes information and participation.

As Calvin Schrag states, much of philosophy has struggled over the fact/meaning issue. Yet both are important. We need to understand information within a "fabric of fact" (Schrag, 1986, pp. 175-190). The "fabric of fact" suggests that the historical, social, cultural, and personal biases of the recorder of information and the recipient of information make all the difference. A story weaves facts together into a coherent fabric of meaning.

Walter Fisher (1987) offers criteria for testing the adequacy of "weaving" of a narrative. First, a story needs to have internal coherence; it needs to be consistent over time and "hang together" to form a clear picture. Second, a story needs to make sense for the community – it needs fidelity to the community's values as the story meets the historical moment. In the words of Vico (cited in Gadamer, 1986), a narrative must generate common sense—it must fit the historical moment in a way such that the community can verify its truthfulness. Danto (1985) connects the notions of story and historicality: historicality provides the context for understanding a given story. A story does not include all events; it glues some events together, offering significance for an event or action. Whether one's story meets historical demands appropriately is tested by common sense for a given community and fidelity with the genuine needs of a given historical moment.

A manager meets the demands of the historical moment and the audience – employees and superiors. Secular and private institutions have different virtue structures; both have value in a postmodern age (Engelhardt, 1990, pp. 41-42) and must articulate appropriate moral visions to their audiences or risk losing fidelity. For instance, a manager using religious language to explain all the success of a secular company is likely to miss the historical moment and the needs of the people. However, discussing the virtues of group work, casting a vision that meets the needs of the market begins to frame an appropriate story about successful response to the historical moment. On the other hand, religious language might be appropriate for a Catholic health care institution seeking to find new ways to live the essential vision and mission of the Catholic Church in a dynamic environment (Doughterty, 1999; Giganti, 2001). Only by making its story clear can such an institution respond well to social, political, and economic changes of a given historical moment (Cochran, 1999).

This essay assumes that postmodernity and an information age point to the need for a *story* as a *philosophical sense of home* as the third meaning metaphor (after *place* and *self*) to guide us.   This basic assumption accepts five observations about the introduction of story framed in this manner into organizational life: 1) we cannot assume that an organization knows its story beyond the bottom line of fiscal health; 2) a story is more than a mission written on paper; 3) an organizational story is told for manifold reasons, including, but not tied exclusively to, productivity, 4) not every story will be constructive; we need to differentiate good and bad stories; and 5) despite multivocality of "stories" in an organization's culture (Boje, 1995), a publicly recognized and articulated story needs to provide the standard or common center for the organization. Story-laden understanding is a way of offering temporal clarity for an organization and its people, moving information to meaning.

Limited stories. When a culture loses a metanarrative, people discover incommensurability in decision-making; such moments leave organizations bereft of leadership with story-based consensus.   When a culture loses its way (narrative direction), materialism triumphs. When a person loses the way, individualism (de Tocqueville, 1966, pp. 395-397) triumphs, and materialism and individualism substitute for story-framed meaning.   When an organization does not have a clear story, the bottom line of profit (quarterly reports), the organizational equivalent of individualism (Dell'Oro, 2001), drives interaction.   Individualism and materialism miss the larger mission/story of the organization. Schrag's "fabric of fact", the context, the meaning center, is missing.   Both for-profit and not-for-profit organizations are vulnerable to individualism and bottom line considerations if a more comprehensive organizational story is not articulated or goes unnoticed.

With family members located in diverse geographical locations, mobility and lack of narrative connection result in what Berger, Berger, and Kellner (1973) called "the homeless mind". Individualism ceases to ground or connect a person to values, mores and institutions larger than the self. The communicative background of *us* is simply lost. Individualism attempts to function without a sense of background that offers guidance, *sensus communis*:

Any specific knowledge has *background* (phenomenology calls it a *horizon*). That is, whatever is specifically known assumes a general frame of reference. Also, the discrepant reality definitions of everyday life require some sort of overall organization. In other words, the individual needs overarching reality definitions to give meaning to life as a whole. (Berger, Berger and Kellner, 1973, p. 15)

Bellah and associates (1991) point in a similar direction. They describe the limits of life without some form of institutional commitment beyond the individual. Bellah, et. al. cite anthropologist Mary Douglas: " 'The most profound decisions about justice are not made by individuals as such, but by individuals thinking within and on behalf of institutions.' We can extend her insight by saying that responsibility is something we exercise as individuals but within and on behalf of institutions" (p. 13).

A limited perspective for a person is about "me". A limited perspective for an organization is about "profit" (Dell'Oro, 2001). Just as postmodern scholarship does not disregard agency, but dethrones it, placing it within a fabric of other considerations, the same is true for understanding the importance of profit in an organization. Profit is and should be a significant variable in an organization; it just cannot be the only one (McCrickerd, 2000; Werhane, 2000). Just as the notion of self or agency should assist us when embedded within other issues, such as historicality, socio-cultural considerations, and vocational location, we need to embed the story of profit within multiple considerations (e.g., Bovens, 1998; Magill and Prybil, 2001). A story responsive to organizational responsibility cannot permit organizations to limit discussion to the bottom line alone. Granted, every profit and not-for-profit lives and dies by numbers, but communities die when organizations attend only to profit lines (Maltby, 2001; McCrickerd, 2000), putting at risk the quality of life in the community, a major drawing card for recruiting organizational talent (Coile, 2001). We need organizational stories that offer an expanded understanding of productivity, restraining a "bottom-line only" mentality.

Lasch (1991) describes the common vision of modernity as overconfidence in progress and science, part of the fallen metanarrative of the Enlightenment. In an industrial organization, confidence in continual progress encourages increased expansion and increased use of scientific techniques, ignoring the word *limits*. Lasch (1991) revisits the notion of limits, assuming that every narrative is limited--no one metanarrative can guide us. Notably, as postmodernity makes us more aware of our diversity, it is the lower-class value placed on "limits", not upper-class confidence in progress, that awakens an organization to an era of *limits, the particular,* and *difference.*

Postmodernity calls into question our confidence about the future, permitting discomfort of limits to return, ingredients common in a lower middle class value structure. Of course, too much stress on limits can permit racism, sexism, and fear of the stranger to dominate perspectives. Such is the dark side of limits without awareness; knowledge of limits, however, is essential to our humanity. "Humble narratives" assume limits; they compose background communicative agreements when knowingly engaged. Humble narratives offer an answer to postmodernity's suggestion that no single virtue structure or metanarrative can guide us (MacIntyre, 1984). In an organization, this principle suggests that it is unlikely that people will agree on one story about the organization without a clear public display of what constitutes the story, why it is important, and what communicative social practices give life and congruence to a given story.

The presuppositional base for storytelling acknowledges limits, the particular, and historical sensitivity. We tell stories to differentiate, to limit, and to reveal a unique direction for meaning and action in an organization (e.g., Barry and Elmes, 1997). In other words, we tell stories to offer clarity, guidance, and restraint. Stories differentiate one organization from another (e.g., Coile, 2001, cited in 'Forecasting the health care future,' 2001; Martin, et al., 1983), not just for market niche purposes, but to carve out a sense of historical and professional meaning particular to a given organization (Clark, 1972; Reiser, 1994). Not only should a customer be able to answer, "Why come to this health care institution?"; an employee needs to be able to answer "Why work here and not somewhere else?" (Gerety, 2001). The story needs to offer a reason and a direction for production and for action in the workplace.

Beyond codes and unattended missions. One business person stated that he was tired of codes, mission statements, and strategic plans. Much time was wasted constructing plans that were quickly forgotten, particularly if the personnel changed. A story-driven, "narrative" organization begins with the story/mission of the organization, not a strategic plan. Strategic plans too often offer *a priori* direction before understanding the historical moment. There is no technique for discerning the correct direction; an organization needs to be nimble – flexible and capable of moving with historical changes without losing its way.

The difference between a technique- and a story-framed organization rests within the work of Anatol Rapoport, who differentiated "strategic" and "conscience" oriented thinking (1967). Conscience-oriented thinking involves a value system/story sensitive to the particular needs of a given historical situation. The conscience-oriented thinker works from a story that suggests a value that is manifested in dialogue with the particular needs of the historical moment.

A strategic thinker, on the other hand, wants to discern answers before meeting the particular uniqueness of a given problem. This form of thinking is paradigmatically bound to *a priori* implementation plans. A conscience-oriented thinker connects to a story in constant dialogue with the uniqueness of the given moment. The core values of the story guide, not a technique (e.g, Deal and Kennedy, 1982; Peters and Waterman, 1982). A story-driven organization does not rush to strategic plans but to honest assessment of the needs of the historical moment. The story/mission is the long-term plan. The historical moment offers new opportunities for implementation of the story/mission. The task of story-guided management is first to know the story well, then to respond to the historical moment, and finally to discern necessary communicative social practices connecting the story and the unique needs of a given historical situation.

Story-guided excellence. A story/mission reminds people of an organization's responsibilities to the public, the community, the profession, and to one another

(Bovens, 1998). Organizations are a major part of our lives, a part of the communities in which we live. Bellah and colleagues, in the *Good Society* (1991), suggest that we have gone too far trying to live life without institutional connections and loyalty. Much of the meaning of our lives comes from organizational and institutional involvement.

Gadamer discusses the connection between excellence and productivity in the following fashion. First, there is a community of *sensus communis* (Gadamer, 1986) from which taste for excellence emerges. A story of excellence rests within limits of the community, the customers, the story of the organization, and the unique demands of the historical moment. One detects excellence from a perceptual frame propelled by a "humble" or "limited" story of excellence; our knowledge of limits keeps us looking beyond ourselves. A story is not an individual act. It represents a community and the historical situation, tied to excellence in production that involves multiple characters, offering a sense of meaning in the workplace.

Organizations engage in productivity and excellence to survive. However, any organization needs a center other than profit to distinguish itself. Profit is necessary for every organization, offering little uniqueness or character for the organization. An organization "brand" cannot emerge from profit alone, nor can organizational identity appear out of convenience (agrarian age of *place*) or through relational connection (industrial age of *self*). Neither convenience nor relational connection can "brand" an organization uniquely. What begins and maintains an interest in product "brand" is a story of excellence in which an audience or consumers see congruence between speech and action of marketing and product or service quality. Indeed, research suggests that alignment of organizational activities with a mission is associated with performance success, though managers in not-for-profit health care institutions do not seek aggressively to foster such alignment (Bart and Tabone, 1998).

If the sad events of Enron offer any message, perhaps it is that organizations need a story of excellence and productivity that goes beyond "me", the self, the individual, and connects one to a larger world view. A responsible mission/story takes us beyond concern for "me", beyond profit alone, and connects us to the larger historical moment that recognizes our embeddedness within a culture and society.

Stories as good and bad. Just as there are good and bad people, countries, and families, there are good and bad stories to guide organizations. Dietrich Bonhoeffer (1954), a major community theologian of the twentieth century, stated that some organizations are so vile they need to be burst asunder. There are good and bad stories vying for the chance to guide an organization. The test of stories is in the connection between what is stated and what is practiced. Bok (1979) keys in on public veracity that is tested in action as a truth-telling test. Gadamer (1980) points in a similar direction as he connects *logos* and *ergon* (pp. 1-20), word and deed, in cooperative congruence. The test of story requires it to meet the light of day in a public telling. The task of a manager as storyteller is to tell a story that has public internal and public external fidelity and consistency. The story needs to make sense inside the organization and to outside stakeholders, remaining true to the mission and identity of the organization over time. The story also has to have veracity, its words followed by the deeds of congruent managerial action (Bart and Tabone, 1998; Fritz, Arnett, and Conkel, 1999).

This essay suggests that for all organizations, and particularly organizations with a service commitment, the manager must assume the role of storyteller in a world of competing narrative and virtue structures (MacIntyre, 1984, p. 8). However, before an organization can discern its own story, it needs to have a level of respect to keep conversation going (Rorty, 1979). This essay proposes professional civility as a background story necessary for organizations to maintain a sense of integrity, identity, and ethos--a story beyond the profit motive and oneself alone.

4. A STORY OF PROFESSIONAL CIVILITY

To take us beyond concern for profit and "me" alone is to battle a culture of cynicism. Cynicism finds life within actions that demonstrate little concern for the Other. Cynicism finds life when there is no longer trust that welfare of persons beyond profit and self guide organizational life. Trust battles cynicism with consistency, predictability, and concern beyond profit and "me". This essay suggests that cynicism is countered by an organizational narrative people can trust. An organizational narrative is implemented by a "background" story this essay calls 'professional civility'. A story of professional civility functions as a "humble narrative", providing a dependable communicative background that unites speech and action. A story of professional civility provides the mission or story of an organization with a means of implementation. Professional civility is an implementation story for an organization's narrative.

Professional civility as a background story for implementation of an organizational story offers guidelines for communicative life, framing a reason for productivity of "us", not just "me". Professional civility is a story offered by a storyteller, by management working to improve the larger fabric of productivity within an organization with a clear mission and narrative.

One note of caution – if professional civility is employed as an extension of "human relations" management, then it will fail. The story of professional civility begins not with the individual, but with a narrative framing of the workplace. Professional civility cannot assure psychological happiness or friendship. The key is the organizational story of which workers and management are part; a story unifies fragmented parts of the organization and persons within a common project.

Professional civility is a story that guides behavior within an organizational narrative, foregoing the temptation to focus on relational discourse within an organization, choosing instead to keep attention upon labor and productivity that propel an organization. An organization that understands its productive task and stays on target is likely to be a long-term productive organization; ironically, out of task-centered action, relational connection and a sense of community might emerge (e.g., Gerety, 2001). Professional civility keeps us focused on agreed upon productive tasks, permitting relational connections to prosper as a by-product of working on a common task within an agreed-upon organizational narrative/mission.

4.1 Historical Significance – Why Professional Civility?

The need for professional civility manifests itself in publications such as Stephen Carter's (1998) *Civility: Manners, Morals, and the Etiquette of Democracy.* Professions ranging from law to education have registered concern: "Civility between opposing counsel has deteriorated to the level that it is being critiqued by judges and lawyers locally, regionally and nationally" (Otorowski, 1993, p. 23-26). Florida has generated guidelines for professional civility in the classroom: ". . . honesty, candor, integrity, fair play, courtesy, and respect for others" (Ehrlich and Makar, 1994, p. 18). Publications on incivility in the workplace (Andersson and Pearson, 1999) and "dialogic civility" (Arnett and Arneson, 1999) highlight similar concerns. Professional organizations understand the pragmatic need for a story of professional civility to enhance trust and productivity in organizational life.

4.2 Reclaiming Organizations as Public Spaces.

Adam Seligman (1992) suggests that we reclaim public space by shoring up both the public and the private domains, not collapsing the two. Public and private domains interact dialectically, but are fundamentally different. The public domain requires a different form of constraint than does the private domain, but offers more trust for outsiders in an era of diversity.   A professionally civil organization works to keep a background of agreement on the importance of a public domain, even when workers find each other "troublesome" (Fritz, 2002). Such an organization can function with problematic employees on a personal level if the public domain is professionally supportive and clear. Sociability, or the ability to get along with others in a civil and polite manner, became the hallmark of public discourse in locations such as workplaces that demanded interaction with strangers during the 19th century (Gurstein, 1996).

The relationship of public to private life has engaged increased attention over the past decades (e.g. Arnett and Arneson, 1999; Rawlins, 1998; Sennett, 1974). Several authors note that the historical shift of production from the home to industry (e.g., Gurstein, 1996; Marks, 1994; Scollon and Scollon, 1995) created a dichotomy between the world of the home and public life. One's home became a haven of privacy to which one escaped, avoiding the impersonality of the public sphere; the private domain offered a "haven in a heartless world" (Lasch, 1977).

After World War II, a therapeutic approach to communication stressed self-disclosure and authenticity within nontherapeutic social contexts (Arnett, 1997b). An ideology of openness and self-disclosure closely connected to a push for equality and democracy in all spheres of life, which began at the end of the 19th century and continues today (Gurstein, 1996), permeated the United States' culture. Fading concerns for adherence to traditional social practices and eschewing of formal prescriptions for public behavior created a loss of shared expectations for appropriate interaction in civic/public spaces, leading to emotivist (MacIntyre, 1984), or individualized, standards for corporate and civic behavior. This cultural shift influenced expectations for behavior in organizations, resulting in increased attention to issues of civility and incivility in the workplace: organizational citizenship (Organ, 1988), organizational deviance (Robinson and Bennett, 1995), trust in organizations (McAllister, 1995), unpleasant work relationships (Fritz, 2002), and ethical behavior in the workplace (Knouse and Jiacalone, 1992; Robertson, 1993), including ethical incommensurability (Primeaux, 1992) and the need for publicly articulated codes of ethical conduct (Fritz, Arnett, and Conkel, 1999). The meaning of public and private in organizational life changed as the historical moment shifted.

Theorizing by Marks (1994) explicitly illustrates the turn toward individualization or privatization of the public sphere. Marks argues that the differentiation created by segmented organizational life fashions the public world into a private space, creating the potential for intimacy. Marks suggests a reconceptualization of privacy and its relationship to intimacy and to social ecologies. He argues that we should link private life not to a particular context or social ecology but to individuals' "constructive inclinations and opportunities" (p. 854) – that is, both the private and the public become individualized. One constructs a private world from one's interests and activities. One then inhabits that private world, which may cross boundaries of traditionally distinct contexts – work, family, recreation, for example. Self-disclosure becomes organizationally key, permitting multiple persons to enter one's private world.

Studies of various organizational contexts address private concerns that surface in the public arena of work, resonating with Marks's call to redraw the boundaries of private worlds. These studies highlight implications and consequences of the confluence of these two spheres. In some cases, we see that the private becomes an issue of public concern because of its effect, intended or not, on the corporate good. This potential threat to organizational health often prompts organizational leadership to offer programs to address personal problems (e.g., Employee Assistance Programs to help employees overcome drug or drinking problems (Hollinger, 1988)). In other cases, the negative effect on the corporate good resulting from the private concern is less discernable, for instance, the problem of employee weight control stemming not from corporate, but individual, concern (e.g., Colvin, Zopf, and Meyers, 1983).

Some of these studies point out the potential danger of intimate spheres in public places, of work sustained by relational ties, bonds which create a context for supporting or enabling behavior with damaging corporate and individual consequences (e.g., Hollinger, 1988; Roman, Blum, and Martin, 1992). They seem to understand what Parker Palmer (1989) called the "tyranny of intimacy". Intimacy in the workplace invites keeping another under watch. Anyone from a small town can relate to this misuse of the private under the guise of a constructive social good. Ray (1993) points us to similar dangers as supportive communication relationships take on political, informational, codependent, and hegemonic features.

One example of the surfacing of private issues within the public realm is alcohol and drug use in the workplace, which affects workplace productivity and "sunk costs" of employee training and socialization. Hollinger (1988) reports that employees who socialized frequently with coworkers were twice as likely as their nonsocializing counterparts to come to work intoxicated, regardless of their age or gender. Manello and Seaman (1979, cited in Hollinger, 1988) suggest that workers' inability to distinguish social boundaries between work and nonwork situations may exacerbate inappropriate behavior.

This inability to distinguish boundaries may stem from informal work cultures that encourage friendships that transcend task-focused organizational roles, extending to extensive socializing outside of work. " . . . [W]orkers who interact with one another away from the job site will . . . be more tolerant of any deviance by their peers" (Hollinger, 1988, p. 445). In other words, the culture of friendship becomes the foreground, trumping productivity and role-governed interaction at work. Roman et. al.'s (1992) study confirms the influence of workplace interaction on alcoholic employees by pinpointing a "workplace enabling effect of coworker and supervisory support in getting one's job done" (p. 284). Their study illustrates the danger of moving too far toward the pole of the private. A story of professional civility moves one back to the public domain and frames a common center for the organization.

4.3 A Public Mission Around a Common Center.

In this temporal moment the key to a civil organization begins with a clear public narrative supported by a strong ethos that recognizes something beyond "me". In this moment of reclaiming organizational space as a civic public domain, one must continue to contend with a "therapeutic culture" (Rieff, 1966/1987). A therapeutic culture begins with "self"; a civic culture begins with the public narrative that guides a given organization or community. As Stephen Carter (1998) reminds us:

> I have always enjoyed cemetery walks. A cemetery reminds us of the impermanence of mortal life, suggesting a need for dispatch in fulfilling our moral obligations; but it also reminds us of the durability of faith, which suggests that these obligations themselves are transcendent, not contingent. Like the Alamo, like a national park, like a moment of silent prayer, a cemetery helps us to find the peace within which we remember that we do not belong entirely to ourselves. (p. 292)

Stories of sacrifice remind us of life beyond ourselves; in much less dramatic fashion the story of professional civility reminds that there is more than "me". A story of professional civility situates us within a fabric of persons, events, ideas, plans, within the common center of the organization, its identity, its ethos. A manager's task is to work with employees within the organization's mission and narrative, keeping the conversation going with a baseline agreement of civility toward one another, even toward those who are personally disliked, in order to maintain the institution's integrity for external and internal audiences. A mission maintains its credibility through its social practices; professional civility is a story of implementation of an organization's mission through social practices: how do organizational members work together around this common center?

Professionals need to show respect for both the persons with whom they work and the vocation that calls them to work each day. The same story that informs one about what it means to be a professional in a given area—whether health care, teaching, or engineering—must be as well a story of respect for co-workers, superiors, subordinates, customers/clients/patients, and for the local home, the local organization. Professional civility begins with threefold respect: the craft or profession, persons at work, and the local organization.

Managers need to bring two stories together: the mission/narrative of the organization and the story of professional civility. These two stories combine in a holistic fashion to generate respect on three levels: respect for the profession, for the people, and for the organization guides one's communicative style. Within different "local homes", particular practices that show respect will differ to some degree, being derived from that organization's narrative. For example, two organizations may engage conflict differently (e.g., directly as opposed to indirectly), in which a particular style may be appropriate for that organizational culture, and thereby considered professionally civil in that context. Professional civility will be manifested in different particular practices in different places while still adhering to general guiding principles, discussed next.

5. PRACTICES OF PROFESSIONAL CIVILITY

Communication is both the carrier and the container of a "community of memory" (Bellah et. al., 1985, pp. 152-155) in professional organizations. How we talk about our tasks and responsibilities in an organization frames the *communicative background* or *story* for organizational life. We empower persons within the organization as *storytellers* as we listen and as we recognize connections between discourse and action. Managerial *practices* underscore the *veracity* of the story, providing a legitimate rationale for cynical protection from rhetoric when incongruent action emerges. Professional civility at the managerial level requires congruence between the rhetoric of proclamation and the rhetoric of action. Managers as *storytellers*/organizational leaders bear the burden of doing what the story/mission proclaims and providing a context where the organizational narrative can be practiced by others.

A communicative ethic in the form of a story of professional civility provides background support for implementing communicative social practices that verify, support, reinforce, and give life to an organization's narrative. The presence or absence of these practices determine whether a communicative background of professional civility responsive to an organization's narrative or a background of cynicism will guide an organizational culture. The following material lists communicative social practices that invite and support a background communication ethic or story of professional civility.

5.1 Public Action Around a Clear, Limited Story

5.1.1 Understand and Communicate the Common Center

Professional civility moves from focus upon the individual to a common center, the mission/story of the organization. The human relations movement in organizational theory was a "moral cul-de-sac" (Arnett, 1997b). Not only were its assumptions wrong (satisfaction does not lead to greater productivity (see C. Fisher (1980)), but as the historical moment shifted, continued focus on the self in organizational settings created unmet high expectations, leading to cynicism. Organizations cannot meet al$l$ human needs. People need lives outside of work.

This principle carries over to organizational rewards. Imagine that Ms. Jones receives an opportunity to receive more training in her job area. Unlike a human relations approach, professional civility reminds one that the opportunity arises from Ms. Jones's productivity and out of a desire to promote the organization, not because Ms. Jones is special or must have her needs met to be productive. As a *byproduct* of productivity and organization assistance, employees may flourish. Even for an organizations with a strong "person mission" (Hebden, 1986), assistance to the person must be a by-product of attention to an organizational common center.

Critical management studies (e.g., Alvesson and Willmott, 1992) critique power interests, reminding us that all action is done in someone's interest. Professional civility is a story that assists organizations in making their interests public, taking "interest" out of the domain of "unobtrusive control" (Barker and Cheney, 1994), publicly admitting goals and aims. Professional civility does not attack contemporary Western capitalism and traditional organizational forms; this essay agrees with Ashcraft (2000): these forms are likely to be with us for some time (p. 380). Nonetheless, it is essential that profitable organizations accomplish other socially worthy goals, as Werhane (2000) points out, citing the work of Collins and Porras (1994) on visionary organizations. Organizations can maintain their integrity, though their actions will not please all people, especially when institutions have a strong moral story (e.g., Uttley, 2000).

5.1.2 Encourage Public Standards for Behavior

Incivility in organizational settings is on the rise (Andersson and Pearson, 1999). One reason for "spiraling incivility" may be, paradoxically, the creation and fostering of informal organizational cultures. In today's world, we cannot assume common standards of morality and must accept incommensurability (MacIntyre, 1984). In organizations, public standards provide clarity and direction in a public setting.

5.1.3 Avoid Blurring the Boundaries Between Public and Private Spaces, Relationships, and Discourse (Arnett, 2001)
Though the bifurcation between public and private spheres instituted by Weberian bureaucracy has been critiqued, especially by feminist authors, recent theorizing recognizes that in a moment of disputed virtue structures, the public domain may be necessary once again (Ashcraft, 2000). Moving issues of the private into the public can create hostile work environments and potential incivility due to employees' unrealistic expectations for social need satisfaction at work. Human beings need social stimulation, and most of us would like to work with pleasant people. However, people appear to have different needs for close friendships at work (e.g., Marks, 1994). Even people who seek close friendships on the job and find them would do well to be cautious. Research on blended relationships (Bridge and Baxter, 1992) suggests that friendships with work associates carry challenges associated with differing role expectations of friends and coworkers. For instance, Ashcraft (2000) points to concerns related to favoritism projected onto coworkers who are friends or dating partners. When expectations for private relationship norms trump requirements of role relationships, task productivity suffers. When information is used as a means to foster intimacy, the need for intimacy overrides the need to be judicious with organizational information. When the ties that bind find strength from disclosure of organizational information, trust wanes in an organization.

Relationships at work are inevitable and potentially helpful; one must, however, recognize and respond to appropriateness of relational and organizational roles at particular moments. Making organizational decisions based on friendship patterns rather than what is good for the organization is unwise; it decreases trust. An organizational story of professional civility shapes a culture that rejects "the triumph of the therapeutic" (Rieff, 1966/1987) and the ideology and "tyranny of intimacy" (Palmer, 1989). A story of professional civility challenges employees to celebrate appropriate distance, which makes working with those who are different from oneself possible, providing necessary space for communicating productively with those to whom one is and is not close in the workplace. Making public the expectation that people make decisions based on the good of the whole permits even friends to find freedom at work.

5.1.4 Avoid Inappropriate or Contradictory Organizational Metaphors
As Morgan (1986) suggests, metaphors provide "one-sided insights" (p. 13) into organizational life. The metaphors we use for organizations create and shape understandings of appropriate behavior (McCrickerd, 2000). The metaphor of family, for instance, misleads, inviting an "old boy" or "old girl" network orientation as well as the tyranny of intimacy. Unrealistic expectations for familial ties in an organization can turn to cynicism when "family" members find no promotion or worse.

5.1.5 Work Within Limits
An organizational mission provides argumentative parameters for actions, the visionary limits on organizational action. No organization has limitless material resources; managers are accountable to stakeholders and stockholders for resource use. Individual managers have finite time and energy. To avoid rhetorical overreach, managers should not promise more than they can deliver. Employees recognize the integrity of realistic commitments (Fritz, Arnett, and Conkel, 1999).

5.2 Employees Invited into Narrative Participation

5.2.1 Understand Hiring as an Invitation to Join a Larger Story
The first opportunity to focus on the common center of an organization is during the interview. Managers avoid the temptation to beg employees to join them when a story/mission drives the organization. The temptation to focus on the potential employee is especially strong in an era of scarce employee resources and in an environment still flirting with a "therapeutic culture". Instead of suggesting to a potential employee what an organization can do for that person, a manager needs to tell the organizational story. An organization with a story should both attract and repel potential employees. No organizational story fits all people. Telling an organization's story clarifies its mission and expectations, providing a publicly articulated realistic job preview (Wanous, 1992). Story/mission hiring seeks to avoid rhetorical overreach in interviews, which often leads to future cynicism from unmet high expectations.

5.2.2 Mentor Employees into a Clear Organizational Story/Mission
Formal mentoring programs are on the rise (Tyler, 1998, cited in Ragins, Cotton, and Miller, 2000), but have been touted as less effective than informal mentoring relationships with regard to relational satisfaction (e.g., Nemanick, 2000). Closer examination of research on ineffectiveness in mentoring suggests confusion about the functional purpose of the mentoring. The purpose of mentoring is not employee comfort or satisfaction, but understanding the common center of the narrative inviting one to participation. Formal mentoring moves the focus of attention from personal, relational connection to public accountability.

Mentors can help new employees navigate temptations common to life in a new location. One temptation for new employees in this historical moment is emotivism, decision-making by personal preference (MacIntyre, 1984): "I feel things should be this way." In an age of narrative disagreement, it is easy to fall back on the way "I" think things should be done, confusing personal discomfort with organizational error. A mentor assists with an organization's mission/story and with appreciation of the unique manner in which diverse people contribute to the story. A mentor must stop quickly the emotive impulse to critique anyone doing a task differently than "I" might.

Employees who "watch" other employees and criticize them contribute to an organizational communicate climate – that is, shared perceptions of formal and informal organizational practices (Reichers and Schneider, 1990) — of criticism that can be as harmful as destroying the organization's product. Messages have a powerful influence on perception in organizations (e.g., Sias, 1996; Sias and Jablin, 1995). Managers in a climate rife with employees watching each other soon discover organizational members who are constantly on guard. "Neurotic guilt" results when one has done nothing wrong, but is worried and fatigued from constantly being watched (Buber, 1965, pp. 126-127). Constant criticism and employee "voyeurism" may contribute to neurotic guilt of looking over one's shoulder, wondering what the other now sees.

When employees move to a new organization, they may be tempted to carry the old organization's story with them. They will be tempted to embrace a past provincialism ("we did it this way in the other place"). "Past provincialism" is a type of emotivism with the source of "good" rooted not in the self, but in one's previous location. One fails to consider the narrative of one's current "local home" and looks instead to a past "home". As persons enter new organizational stories, they must break away from old practices and enact new ones. Employees may suffer from seeing the world from the standpoint of a previous location as they change jobs within the same firm. The organizational socialization process happens again for such employees (Kramer, 1993). Likewise, when an organization is re-engineered or merges with another firm, employees encounter a new way of organizational life. The temptation to seek stability by clinging to past meanings requires mentors who can assist new or relocated employees to focus upon the story governing their current "home".

5.2.3 Identify "Standard-Bearers" of the Organization's Mission/Story

"Organizational citizens" (Organ, 1988) go beyond the call of duty in carrying out the organization's mission. These individuals assist others by their excellence. Aristotle, in the *Nichomachean Ethics*, outlines a "standard-bearer" as someone who knows the craft and is perceived by others as a representative of the craft. We see Thomas Jefferson as a standard-bearer for an intellectual President and Martin Luther King, Jr. as a standard-bearer for civil rights. An organization must cultivate standard-bearers, permitting their practices to work as visible, living examples of the organization's narrative.

5.2.4 Marginalize Problematic People

A rule of thumb: It takes five "healthy" people to counteract the influence of one "problematic" one. "Troublesome others" (Fritz, 2002) or "problematic others" (Arnett and Fritz, 2001) take the focus off productivity by behaviors such as bringing personal problems to work, behaving in an unprofessional manner (yelling or screaming), and bossing others around without having the authority to do so (Fritz, 2002). If it is within one's power, one should get these people away from others if they cannot be fired. Another alternative is to turn a problem into an opportunity by letting a problematic other provide experience for managers in training. As always, if a problematic person ceases to be problematic, invite the person into the narrative again. Moving the focus of attention to productivity related to the mission/story of the organization does not assume that we will like everyone in the organization. Professional civility calls for behavior that keeps the focus of attention upon the task at hand, regardless of one's liking for another.

5.2.5 Avoid a Culture of "Emotivistic Opinion"

In a culture characterized by emotivism (MacIntyre, 1984) and expressive individualism (Bellah, et. al., 1985), ideas sometimes do poor battle with opinions. Employees with good ideas require a voice. Ideas are public, calling for public examination. Opinions rest within relational loyalty or organizational power. A mission/story invites ideas, remaining open and flexible, in contrast to an ideology that is fixed and closed (Arnett, 2001). Employees should know that ideas gather attention and a hearing and understand that not all ideas can or should be implemented; such knowledge will decrease cynicism from unmet high expectations.

5.2.6 Employ "Jeffersonian", not "Athenian", Democracy (Arnett, 2001)

Requiring consensus for all decisions is time-consuming and difficult to achieve (e.g., Mansbridge, 1973). Participation can turn to entitlement, a situation in which people become more interested in having their voices heard than in getting the job done or advancing the product. An alternative is to have committees make recommendations for managerial approval. Working through representative committees permits bounded participation focused on productivity. The key is to hear voices, but voices tied to the mission/story, tied to realistic expectations, and tied to one's task.

6. CONCLUSION

A postmodern era offers opportunity for diverse institutions to offer clear, public stories/narratives representing particular moral traditions (Engelhardt, 1990). A health care institution's integrity depends on the living out of its narrative by employees, made possible by framing institutions as locations for public discourse around a common center. A story of professional civility offers managers of health care institutions an opportunity to work with employees in ways that maintain institutional integrity and ethos, encourage productivity, and minimize cynicism and individualism. The rhetorical task of management in a postmodern age is to tell a story that provides a "why" for external and internal audiences, giving them a reason to offer allegiance to a particular health care institution. Removing focus away from the individual self and toward an organization's narrative permits institutional identity and practices to shape an organizational culture and to offer a sense of organizational trust. Trust in an organization's story requires that a manager do what is offered in words. Congruence of word and deed determines the difference between a culture of cynicism and individualism, resulting in destruction of institutional integrity and ethos, or one of productivity and trust, resulting in health and maintenance of institutional integrity and ethos.

Duquesne University
Pittsburgh, Pennsylvania, USA

BIBLIOGRAPHY

Alvesson, M., & Willmott, H. (Eds.) (1992). *Critical Management Studies.* Thousand Oaks, CA: Sage.

Andersson, L. M., & Pearson, C.M. (1999). 'Tit for tat? The spiraling effect of incivility in the workplace.' *Academy of Management Review, 24,* 452-471.

Arendt, H. (1958 / 1989). *The Human Condition.* Chicago: University of Chicago Press.

Aristotle. (1985). *Nichomachean Ethics,* Irwin, T. (trans.). Indianapolis, IN: Hackett.

Arnett, R. C. (1997a). 'Communication and community in an age of diversity.' In: Makau, J., & Arnett, R. C. (ed.), *Communication Ethics in an Age of Diversity* (pp. 27-47). Urbana: University of Illinois.

Arnett, R. C. (1997b). 'Therapeutic communication: A moral cul-de-sac.' In: Longnecker, S.L. (ed.), *The Dilemma of Anabaptist Piety: Strengthening or Straining the Bonds of Community* (pp. 149-159). Camden, ME: Penobscot Press.

Arnett, R. C. (2001). 'Dialogic civility as pragmatic ethical praxis: An interpersonal metaphor for the public domain.' *Communication Theory, 11,* 315-338.

Arnett, R.C. & Arneson, P. (1999). *Dialogic Civility in a Cynical Age.* Albany: State University of New York Press.

Arnett, R.C. & Fritz, J.M.H. (2001). 'Communication and professional civility as a basic service course: Dialogic praxis between departments and situated in an academic home.' *Basic Communication Course Annual 13,* 174-206.

Ashcraft, K.L. (2000). 'Empowering professional relationships: Organizational communication meets feminist practice.' *Management Communication Quarterly 13,* 347-392.

Barker, J.R., & Cheney, G. (1994). 'The concept and practices of discipline incontemporary organizational life.' *Communication Monographs 61,* 19-43.

Barry, D. & Elmes, M. (1997). 'Strategy retold: Toward a narrative view of strategic discourse.' *Academy of Management Review 22,* 429-452.

Bart, C.K. & Tabone, J. (1998). 'Mission statement rationales and organizational alignment in the not-for-profit health care sector.' *Health Care Management Review 23,* 54-69.

Bellah, R., Madsen, H., Sullivan, W., Swidler, A. & Tipton, S. (1985). *Habits of the Heart: Individualism and Commitment in American Life.* Berkeley: University of Berkeley Press.

Bellah, R., Madsen, H., Sullivan, W., Swidler, A. & Tipton, S. (1991). *The Good Society.* New York: Alfred A. Knopf.

Berger, P., Berger, B. & Kellner, H. (1973). *The Homeless Mind: Modernization and Consciousness.* New York: Random House.

Best, S. & Kellner, D. (1991). *Postmodern Theory: Critical Interrogations.* New York: The Guildford Press.

Bisson, D. (2002). 'Institutional integrity: Values in action.' *Health Progress, 83(2),* 10-13.

Boje, D. (1991). 'The storytelling organization: A study of storytelling performance in an office supply firm.' *Administrative Science Quarterly, 36,* 106-126.

Boje, D.M. (1995). 'Stories of the storytelling organization: A postmodern analysis of Disney as "Tamara-land."' *Academy of Management Journal, 38,* 997-1035.

Bok, S. (1979). *Lying: Moral Choice in Public and Private Life.* New York: Vintage Books.

Bonhoeffer, D. (1954). *Life Together.* New York: Harper & Row.

Bovens, M. (1998). *The Quest for Responsibility: Accountability and Citizenship in Complex Organizations.* NY: Cambridge University Press.

Bridge, K. & Baxter, L. (1992). 'Blended relationships: Friends as work associates.' *Western Journal of Communication, 56,* 200-225.

Bruhn, J.G. (2001). 'Being good and doing good: The culture of professionalism in the health professions.' *Health Care Manager, 19(4),* 47-58.

Buber, M. (1958). *Paths in Utopia.* Boston: Beacon Press.

Buber, M. (1965). 'The knowledge of man: A philosophy of the interhuman.' New York: Harper & Row.

Burke, R.J. (2002). 'The ripple effect.' *Nursing Management, 33(2)*, 41-42.

Carter, S.L. (1998). *Civility: Manners, Morals, and the Etiquette of Democracy*. New York: Basic Books.

Clark, B.R. (1972). 'The organizational saga in higher education.' *Administrative Science Quarterly, 17*, 178-184.

Cochran, C.E. (1999). 'Institutional identity; sacramental potential: Catholic health care at century's end.' *Christian Bioethics, 5*, 26-43.

Coile, R.C., Jr. (2001). 'Magnet hospitals use culture, not wages, to solve nursing shortage.' *Journal of Healthcare Management, 46*, 224-227.

Colvin, R.H., Zopf, K.J., & Myers, J.H. (1983). 'Weight control among coworkers: Effects of monetary contingencies and social milieu.' *Behavior Modification, 7*, 64-75.

Danto, A. (1985). *Narration and Knowledge*. New York: Columbia University Press.

Deal, T.E. & Kennedy, A. A. (1982). *Corporate Cultures: The Rites and Rituals of Corporate Life*. Reading, MA: Addison-Wesley.

Dean, J.W., Brandes, P. & Dharwadkar, R. (1998). 'Organizational cynicism.' *Academy of Management Review, 23*, 341-352.

Dell'Oro, R. (2001). '"Integrity" and "compliance".' *Health Progress, 82(5)*, 29-36.

De Tocqueville, A. (1966). *Democracy in America*. New York: Knopf.

Dougherty, C.J. (1999). 'On the road to Jericho.' *Christian Bioethics, 5*, 66-74.

Ehrlich, R., & Makar, S.D. (1994). 'Professional civility and aspirational conduct.' *The Florida Bar Journal, 66(3)*, 14-20.

Engelhardt, H.T., Jr. (1990). 'Integrity, humaneness, and institutions in secular pluralist societies.' In: Bulger, R.E. & Reiser, S.J. (eds.), *Integrity in Health Care Institutions: Humane Environments for Teaching, Iinquiry, and Healing* (pp. 33-43). Iowa City: Iowa University Press.

Fisher, C. (1980). 'On the dubious wisdom of expecting job satisfaction to correlate with performance.' *Academy of Management Review, 5*, 607-612.

Fisher, W. (1987). *Human Communication as Narration: Toward a Philosophy of Reason, Value, and Action*. Columbia: University of South Carolina Press.

'Forecasting the health care future.' (2001). *Marketing Health Services, 21(3)*, 16-22.

Fritz, J.M.H. (2002). 'How do I dislike thee? Let me count the ways: Constructing impressions of troublesome others at work.' *Management Communication Quarterly, 15*, 410-438.

Fritz, J., Arnett, R. & Conkel, M. (1999). 'Organizational ethical standards and organizational commitment.' *Journal of Business Ethics, 20*, 289-299.

Gadamer, H. (1980). *Dialogue and Dialectic: Eight Hermeneutical Studies on Plato*. New Haven: Yale University Press.

Gadamer, H. (1986). *Truth and Method*. New York: Crossroads.

Gerety, J. (2001). 'The spirit of the place.' *Health Progress, 82*, 55-57.

Giganti, E. (2001). 'Living our promises, acting on faith.' *Health Progress, 82*, 32-35.

Green, L.W. (1990). 'The revival of community and the public obligation of academic health centers.' In: Bulger, R. E., & Reiser, S. J. (eds.), *Integrity in Health Care Institutions: Humane Environments for Teaching, Inquiry, and Healing* (pp. 148-164). Iowa City: Iowa University Press.

Gurstein, R. (1996). *The Repeal of Reticence*. New York: Hill and Wang.

Hebden, J.E. (1986). 'Adopting an organization's culture: The socialization of graduate trainees.' *Organizational Dynamics, 15*, 54-72.

Hollinger, R.C. (1988). 'Working under the influence (WUI): Correlates of emloyees' use of alcohol and other drugs.' *The Journal of Applied Behavioral Science, 24*, 439-454.

Kanter, D.L. & Mirvis, P.H. (1989). *The Cynical Americans: Living and Working in an Age of Discontent and Disillusionment*. San Francisco: Jossey-Bass.

Kimball, B. (1992). *The "True Professional Ideal" in America: A History*. Cambridge, MA: Blackwell.

Knouse, S.B. & Giacalone, R.A. (1992). 'Ethical decision-making in business: Behavioral issues and concerns.' *Journal of Business Ethics, 11*, 369-377.

Kramer, M. (1993). 'Communication after job transfers: Social exchange processes in learning new roles.' *Human Communication Research, 20*, 147-174.

Kristoff, A.L. (1996). 'Person-organization fit: An integrative review of its conceptualizations, measurement, and implications.' *Personnel Psychology, 49(1)*, 1-48.

Lasch, C. (1977). '*Haven in a Heartless World: The Family Besieged.* New York: W. W. Norton & Co.

Lasch, C. (1984). *The Minimal Self: Psychic Survival in Troubled Times.* New York: W. W. Norton.

Lasch, C. (1991). *The True and Only Heaven: Progress and its Critics.* New York: W. W. Norton.

Levinas, E. (1987). *Time and the Other.* Pittsburgh, Duquesne University Press.

MacIntyre, A. (1984). *After Virtue: A Study in Moral Theory (2nd ed.).* Notre Dame, IN: University of Notre Dame Press.

Magill, G. & Prybil, L. (2001). 'Guidelines for organizational ethics.' *Health Progress, 82(4)*, 12-15.

Maltby, T.A. (2001). 'The dynamics of value.' *Health Progress, 82(5),* 46-52.

Mansbridge, J. (1973). 'Time, emotion, and inequality: Three problems of participatory groups.' *Journal of Applied Behavioral Science, 9*, 351-368.

Marks, S.R. (1994). 'Intimacy in the public realm: The case of co-workers.' *Social Forces, 72*, 843-858.

Martin, J., Feldman, M.S., Hatch, M.J., & Sitkin, S.B. (1983). 'The uniqueness paradox in organizational stories.' *Administrative Science Quarterly, 28*, 438-453.

Marx, K. (1847/1975). *The Poverty of Philosophy.* New York: International Publishers.

McAllister, D.J. (1995). 'Affect- and cognition-based trust as foundations for interpersonal cooperation in organizations.' *Academy of Management Journal, 38*, 24-59.

McCrickerd, J. (2000). 'Metaphors, models and organisational ethics in health care.' *Journal of Medical Ethics, 26*, 340-345.

Morgan, G. (1986). *Images of Organization.* Newbury Park, CA: Sage.

Nemanick, R.C., Jr. (2000). 'Comparing formal and informal mentors: Does type make a difference?' *Academy of Management Executive, 14*, 136-138.

O'Connor, E. (2002). 'Storied business: Typology, intertextuality, and traffic in entrepreneurial narrative.' *Journal of Business Communication, 39*, 36-54.

Organ, D. (1988). *Organizational Citizenship Behavior: The Good Soldier Syndrome.* Lexington, MA: D. C. Heath and Company.

Otorowski, S. (1993 May). 'Civility and rule 11.' *Washington State Bar News*, 23-26.

Palmer, P. (1989). *The Company of Strangers.* New York: Crossroad.

Peters, T.J. & Waterman, R. H., Jr. (1982*). In Search of Excellence: Lessons from America's Best-Run Companies.* New York: Warner Books.

Primeaux, P. (1992). 'Experiential ethics: A blueprint for personal and corporate ethics.' *Journal of Business Ethics, 11*, 779-788.

Rapoport, A. (1967). Strategy and conscience. In: Matson, F.W. & Montagu, A. (eds.), *The human dialogue: Perspectives on communication* (pp. 79-96). New York: The Free Press.

Ragins, B.R., Cotton, J. L. & Miller, J. L. (2000). 'Marginal mentoring: The effects of type of mentor, quality of relationship, and program design on work and career attitudes.' *Academy of Management Journal, 43*, 1177-1194.

Rawlins, W.K. (1998). 'Theorizing public and private domains and practices of communication: Introductory concerns.' *Communication Theory, 8*, 369-380.

Ray, E.B. (1993). 'When the links become chains: Considering dysfunctions of supportive communication in the workplace.' *Communication Monographs, 60*, 106-111.

Reichers, A.E. & Schneider, B. (1990). 'Climate and culture: An evolution of constructs.' In B. Schneider (ed.), *Organizational Climate and Culture* (pp. 5-39). San Francisco: Jossey-Bass.

Reiser, S.J. (1994). 'The ethical life of health care organizations.' *Hastings Center Report, 24(6)*, 28-35.

Rieff, P. (1966/1987). *The Triumph of the Therapeutic: Uses of Faith after Freud.* Chicago: University of Chicago Press.

Robertson, D.C. (1993). 'Empiricism in business ethics: Suggested research directions.' *Journal of Business Ethics, 12*, 585-599.

Robinson, S.L. & Bennett, R.J. (1995). 'A typology of deviant workplace behaviors: A multidimensional scaling study.' *Academy of Management Journal, 38*, 555-572.

Roman, P.M., Blum, T.C. & Martin, J.K. (1992). '"Enabling" of male problem drinkers in work groups.' *British Journal of Addictions, 87*, 275-289.

Rosenau, P.M. (1992). *Post-Modernism and The Social Sciences: Insights, Inroads, and Intrusions.* New Jersey: Princeton University Press.

Rorty, R. (1979). *Philosophy and the Mirror of Nature.* Princeton, NJ: Princeton University Press.

Schrag, C.O. (1986). *Communicative Praxis and the Space of Subjectivity.* Bloomington: Indiana Unviersity Press.

Scollon, R. & Scollon, S. W. (1995). *Intercultural Communication: A Discourse Approach.* Cambridge, MA: Blackwell.

Seligman, A. B. (1992). *The Idea of a Civil Society.* Princeton, NJ: Princeton University Press.

Sennet, P. (1974). *The Fall of Public Man.* New York: Norton.

Sias, P.M. (1996). 'Constructing perceptions of differential treatment: An analysis of coworker discourse.' *Communication Monographs, 63*, 171-187.

Sias, P.M. & Jablin, F.M. (1995). 'Differential superior-subordinate relations, perceptions of fairness, and coworker communication.' *Human Communication Research, 22*, 3-38.

Sinclair, U. (1905/1960). *The Jungle.* New York: Signet.

Snizek, W.E. & Kestel, J.J. (1999). 'Understanding and preventing the premature exodus of mature middle managers from today's corporations.' *Organization Development Journal, 17*, 63-71.

Spence Laschinger, H.K., Finegan, J. & Shamian, J. (2001). 'The impact of workplace empowerment, organizational trust on staff nurses' work satisfaction and organizational commitment.' *Health Care Management Review, 26(3)*, 7-23.

Sullivan, W.M. (1995). *Work and Integrity: The Crisis and Promise of Professionalism in America.* New York: Harper Collins.

Swales, J.M. & Rogers, P.S. (1995). 'Discourse and the projection of corporate culture: The Mission Statement.' *Discourse & Society, 6*, 223-242.

Taylor, F.W. (1911). *The Principles of Scientific Management.* New York: Harper & Brothers.

Toffler, A. & Toffler, H. (1993). *War and Anti-War: Survival at the Dawn of the 21ˢᵗ Century.* Boston: Little, Brown and Company.

Tompkins, P.K. & Cheney, G. (1983). 'Account analysis of organizations: Decision making and identification.' In Putnam, L., & Pacanowsky, S. (Eds.), *Communication and Organizations: An Interpretive Approach* (pp. 123-146). Thousand Oaks, CA: Sage.

Uttley, L.J. (2000). 'How merging religious and secular hospitals can threaten health care services.' *Social Policy, 30*, 4-13.

Wanous, F.P. (1992). *Organizational Entry: Recruitment, Selection, Orientation, and Socialization of Newcomers (2ⁿᵈ ed.).* Reading, MA: Addison-Wesley.

Werhane, P.H. (2000). 'Business ethics, stakeholder theory, and the ethics of healthcare organizations.' *Cambridge Quarterly of Healthcare Ethics, 9*, 169-181.

Wildes, K.W. (1997). 'Institutional identity, integrity, and conscience.' *Kennedy Institute of Ethics Journal, 7*, 413-419.

PATRICIA H. WERHANE

BUSINESS ETHICS, ORGANIZATION ETHICS AND SYSTEMS ETHICS FOR HEALTH CARE[1]

1. JOHN WORTHY

One evening John Worthy, age forty-seven, brought home information about the new health insurance plans his employer, Factory Inc., was offering. Factory had decided to offer a choice of health plans: GoodCare, a managed care plan, or GoodCare Prime, which had a point of service option. Those who wanted to pay more for it themselves could stay with the company's old indemnity insurance.

After dinner, John and his wife Jane looked over the materials and talked about which plan they should choose. John and Jane were in pretty good health, and their younger daughter was rarely sick. Their older girl was covered by the health service at college. GoodCare's HMO had no deductible, only a small copay for office visits, and there were several doctors in their area on the plan's list of providers. They decided to sign up with GoodCare. They filled out the enrollment form and chose a primary care provider whose office would be convenient to get to; John took the papers back to work the next morning.

Two weeks later they received their member cards and handbook from GoodCare, along with a letter encouraging them to make a "get to know you" appointment with the family doctor they'd chosen. There was no copay for the initial visit, but with one thing and another the Worthys didn't get around to scheduling appointments.

One morning, two months later, John Worthy woke up with a headache. When Mrs. Worthy called at 10:30 to say that John was experiencing a rather severe headache that came on suddenly a couple of hours previously, Fran Davis, John's new family doctor, was concerned. Unfortunately, Dr. Davis was in the middle of office hours with a very full waiting room. The physician and the Worthys were still strangers to one another, and neither the Worthys nor Dr. Davis were very familiar with the HMO's rules and procedures, since Dr. Davis's medical group was relatively new to GoodCare too.

A year ago the medical group, weary from battling payers and paperwork, sold their practice to Physician Management Services (PMS). In exchange for the practice's physical assets, a long-term contractual commitment, and a percentage of

Ana Smith Iltis (ed.), Institutional Integrity in Health Care, 73-98.
© *Kluwer Academic Publishers. Printed in the Netherlands.*

gross revenues, PMS negotiates contracts with the many HMOs and other payers for whom their physicians provide services, takes care of all the paperwork, periodically upgrades the practice's computer information system, and keeps the group within its budget. PMS receives a capitated fee from GoodCare for each enrolled patient to cover all primary care, plus outpatient laboratory and x-ray services. The HMO covers specialists and hospital care, including emergency room visits, separately. Physician Management Services pays each physician a salary, but also transfers some of its economic risk to its physicians in the form of bonuses and penalties tied to productivity and utilization. They established protocols to reduce excessive care, and set a five-patients-per-hour productivity standard. Recently, PMS has also been educating physicians about their tendency to send too many patients to the emergency room, a common response among physicians who are capitated for primary care and whose waiting rooms have become crowded with managed care patients who pay little or nothing for office visits. Last year, high ER costs prompted GoodCare to begin assessing financial penalties on primary care providers who overused this resource.

Dr. Davis was aware that if GoodCare decided, after the fact, that John Worthy did not truly require emergency care, the patient or even she as the primary care provider could be financially at risk. Nevertheless, because she was unfamiliar with Mr. Worthy's medical history, overwhelmed at the office, and somewhat concerned by his symptoms, Dr. Davis suggested that Mrs. Worthy take John to the nearest emergency room if his headache really seemed that severe.

Edward R. Post, the emergency room physician, was concerned. When Worthy presented a little before noon, his temperature was normal, and blood pressure was 158/96. Fundoscopic examination showed no indication of elevated intracranial pressure. John said he'd had severe headaches in the past, but none for maybe the last two years. He acknowledged being a long-time heavy smoker and said he seemed to feel some pain behind one eye. During the examination he also said that his neck might feel just a little stiff. Because his headache was so bothersome, it was hard for him to figure out exactly what hurt. Dr. Post could not directly feel any neck rigidity, but believed that given the severity of the headache as John reported it he should have further tests. Post explained to the Worthys that the problem was most likely just a cluster headache, a severe tension headache, or sinusitis; he did not believe that John had meningitis. But he indicated that there was a possibility, although very remote, of early bleeding in the brain, a tumor, or some other much more serious problem. And so he strongly urged John to undergo a CT scan. The Worthys asked whether the expenses would be covered by their health plan. Post replied that this hospital was not affiliated with GoodCare, and so he could not offer any assurance about reimbursement. Nonetheless, he urged the Worthys to stay for the further diagnostic evaluation.

The Worthys talked things over, but decided to leave the hospital because they were afraid the proposed tests would not be covered. Dr. Post couldn't justify insisting that John stay for tests. He suggested that since their primary care provider was busy they consider calling a neurologist in their health plan. The Worthys thought about visiting the in-plan hospital at a considerably greater distance from their home, but opted instead to return home, from where Mrs. Worthy phoned GoodCare.

Connie S. Rogers, customer service representative at GoodCare HMO, was concerned. She explained to the Worthys that if John visited the emergency department or entered the nonplan hospital near his home, and if it was not declared to be a medical emergency, the entire hospital visit could be disallowed. The Worthys recalled Post's suggestion to see a neurologist and asked if the HMO had any available. Rogers replied that GoodCare's neurologist was a Dr. Newman, and gave the Worthys his number. The Worthys phoned, but were told by Dr. Newman's receptionist that they could not have an appointment until they first received an authorization from their primary care provider. Jane Worthy, increasingly frustrated, phoned back to Dr. Davis to secure the necessary authorization. Unfortunately, Davis's phone line was busy, and calls had been temporarily transferred to an answering service. Jane left a message but, unsure when the message would reach Davis, she phoned back to the HMO.

Connie Rogers again answered. Mrs. Worthy, furious almost beyond words, insisted on talking with the president of the HMO. She was becoming panicky that her husband's worsening condition might bode something life-threatening, and was not at all sure that he could tolerate the eighteen-mile drive to the in-plan hospital. Sensing that litigation might be in the offing, Rogers indicated that although the president was not available, she would try to reach GoodCare's medical director and would talk to him on their behalf. She then phoned the medical director and explained the situation as best she could.

Michael Depp, the medical director of GoodCare, was concerned. As soon as he finished talking with Connie Rogers, he phoned Dr. Davis's office and got through promptly. Upon being updated on the Worthys' current plight, Davis indicated she would resolve the problem and phone Depp back with follow-up information. Davis first phoned the hospital emergency department and learned from Dr. Post that John had not appeared to be in an acute emergency condition, and that the Worthys had declined further evaluation for financial reasons. Dr. Davis inferred from subsequent events that John Worthy's condition seemed to be worsening, however, and that prompt referral to a neurologist was medically justified. Dr. Davis was not familiar with Dr. Newman, and did not know whether he was as experienced as other neurologists she knew in town. Nevertheless, she wanted John to be seen expeditiously by someone who could provide all the care he needed. After phoning Newman's office with the necessary referral authorization, Dr. Davis then phoned the Worthys to inform them that they were clear to visit the neurologist.

Finally, Dr. Davis phoned Michael Depp with the promised follow-up and took the opportunity to vent her own frustration. Depp listened patiently and sympathized with her difficulties. He acknowledged that the "Mickey Mouse run-around", as she'd put it, sometimes did pose real inconvenience. But he explained that GoodCare had tried other forms of cost containment for many years with little result. Thus, they installed the primary care case management model, the preferred hospital and specialist list with discounted fees, and their utilization management system.

Patients were still free to visit any provider, and to receive any care they wanted. But the HMO could only assure full coverage for care it deemed necessary and appropriate. When Mrs. Worthy called Dr. Newman's office a second time to set up the authorized appointment, she told the receptionist that her husband had a "really, really bad headache", and couldn't wait until the first available slot next week. Although the receptionist did not completely appreciate the seriousness of the situation, she did juggle out a next-day appointment.

Jane Worthy was concerned. She drove John back to the ER at 11:00 PM, but even before she pulled up to the doorway, his tortured breathing had stopped. Successive attempts by the hospital staff to revive him failed. John Worthy died of a massive brain hemorrhage as his cerebral aneurysm ruptured.[2]

As almost all forms of health care delivery move into organizations, one can no longer focus merely on physician-patient one-on-one relationships or even on one-on-one relationships between a particular patient or professional and the provider organization such as a hospital or other care center. Health care increasingly is delivered to patient populations by healthcare professionals through complex healthcare organizations run by managers and executives who themselves may or may not be professionally trained in medicine. Payers of health care delivery are no longer patients but include insurers, employers, and local, state, and national governments.

Because of these changes, much of the new work in health care ethics focuses the ethics of healthcare provider organizations. (e.g., Spencer, et. al, 2000; Ozar et. al., 2001; Emanuel and Emanuel, 1996; L. Emanuel, 2000; Hall, 2000; Werhane, 2000; Wong, 1998) Still, an analysis of healthcare organizations may not be adequate to describe the present state of the healthcare profession and health care delivery nor deal with the complex normative issues raised by that delivery or delivery system.

In almost all its dimensions health care as it is currently delivered, rationed, or denied in the United States, is imbedded in a complex set of systems and subsystems that entail a complex set of networks of interrelationships. To deal with ethical issues in health care either from a dyadic or even an organizational perspective often belies what is really at issue and thus ignores a number of elements that are related to the issue in question. John Worthy's health care maintenance, payment, and delivery were within a complex network of managed care, employer, insurance, and delivery organizations. Worthy's employer had contracted out the health care of its employees to Physician Management Services, which, in turn, contracted with GoodCare for the healthcare delivery to these and other employees. Worthy was caught in the morass of these relationships. Yet it is difficult to pinpoint which individual or organization was at fault because of the complex structure of the system in which Worthy's health delivery was a part.

This case illustrates why one needs to address ethical issues in health care in systemic as well as in organizational terms, and why one needs to evaluate the normative content of healthcare systems as well as healthcare organizations, an evaluation that requires what the organizational and scientific literature call "systems thinking" or a systems approach.   That will be the dual task of this chapter. In Part 2, I shall first address the relationship of some concepts in business ethics and organizational responsibility to healthcare organizations.   In Part 3, I shall expand this analysis to help us think about health care as it functions in a complex set of systems and subsystems.

I shall conclude that how a system is designed (its structure), what is included or left out (its "boundary-maintaining processes" in Laszlo and Krippner's terms (1998)), and how the value dimensions of each component of the system interact, affect decision-making. This design set has normative outcomes for the system, for those affected by and operating within that system, and even for some traditional ethical issues in health care such as informed consent, patient autonomy, and privacy. John Worthy could have been saved, but, given the systemic restraints, not merely by one individual professional or by his wife.

In what follows, I shall use the term "healthcare organization" to refer to any organization that manages, finances, or delivers health care. "Provider organization" refers specifically to those organizations such as medical centers, clinics, and hospitals that provide care. A Managed Care Organization (MCO) such as a Health Maintenance Organization (HMO) "integrates the financing and delivery of medical care through contracts with selected physicians and hospitals that provide comprehensive health care services to enrolled members for a predetermined monthly premium" (Inglehart, 1994, p. 1167; Wolf 1999, p. 1634).

2. ORGANIZATIONS, BUSINESS ETHICS AND ORGANIZATION ETHICS

Until recently, before managed care, business issues in the healthcare organizations that delivered care, e.g., physician offices, hospitals, clinics, and other healthcare providers, were relatively insulated from clinical issues. Today this separation of powers and of issues is less possible. In the United States we have created a myriad of for-profit and not-for-profit healthcare organizations including managed care organizations such as HMOs that that manage, finance, and often control the delivery of health care.   Payment to professionals and provider organizations is ordinarily through insurers or government agencies, not by patients, and often the extent and quality of reimbursed care is measured through capitation, rationing, diagnostic related groups (DRGs) set by the insurers, or, in the case of the state of Oregon, by legislators (Spencer, et. all, 2000).

In the contemporary healthcare setting, financial, clinical, and professional issues are all so interrelated that one cannot neatly separate out, say, the cost of an MRI from a patient's need for it, from the professional expertise that determines the desirability of that protocol, nor from the HMO guidelines that specify when and at what cost it can be administered.   Thus there is a need to take into account the interrelationships between clinical, patient, professional, management, and financial aspects of those organizations as these are managed by or in HMOs.

To make sense of the organizational and systemic dimensions of healthcare delivery and payment, most of which is through various organizations, we must first think about whether and in what ways we can hold organizations as well as individuals morally responsible. Some healthcare organizations are corporations, which are treated as legal persons under the law. Others have other forms of organizational structure. But whether or not they are treated as *legal* persons, it is important to determine whether and in what ways organizations are moral agents and thus morally responsible for their actions.

2.1 Can We Hold Organizations Morally Responsible?[3]

There are a number of senses in which we ordinarily think of organizations as moral agents. Like individuals, organizations set goals.[4] These goals are often defined in a mission statement, delineated in policies, or operationalized in the organization or organizational culture and activities in which it is engaged. In ordinary language we refer to organizations as actors, and we hold them, like individuals, responsible. For example, we say that the University of Virginia Medical Center was responsible for the now infamous baby switch in 1995 when two babies were sent home to the wrong parents, a switch that was not discovered until three years after the birth of the babies (Shear, 1998, p. B-1). We praise Catholic Healthcare West for its new Mistakes Project, a series of initiatives aimed at reducing medical practice errors (Bayley, 2001). We hold these organizations, like individuals, accountable for their actions. The University of Virginia Medical Center, not simply its healthcare professionals and staff, was held morally as well as legally liable for the baby switch. In the John Worthy case we hold GoodCare as well as its administrators and contracted professionals and delivery organizations responsible for John Worthy's death.

Organizations sometimes act as single units and exhibit intentional behavior. (Werhane, 1985) Many organizations appear to "think about" their desires, beliefs and goals, and some organizations or persons acting on behalf of corporations seem to engage in moral deliberation and self-analysis as well. For example, during the silicone breast implant controversy Dow Corning, the leading manufacturer of silicone breast implants, debated about the morality and feasibility of manufacturing implants. Of course, Dow Corning the firm did not literally engage in self-reflection and moral self-analysis about breast implant manufacture, rather its constituents (the board, managers, employees and legal agents) did so on behalf of the corporation. Dow Corning's so-called "corporate actions" were a result of these deliberations that subsequently directed actions of persons who functioned as agents for the corporation. To continue the example, Dow Corning was responsible for the manufacture of silicone breast implants. But the corporation, in fact, did not design, manufacture or market implants; engineers and other employees working at Dow Corning did so. Still, we hold Dow Corning responsible because the persons who made implants were acting as agents on behalf of the corporation. We then say that the corporation performed intentional actions. But literally the actions were a result of the activities of a collection of persons or groups operating on behalf of Dow Corning. Thus the intricate web of intentional behavior exhibited in organizational decision-making procedures coupled with resulting actions by agents on behalf of the institution produce collective, organizational "actions".

Organizations also engage in reciprocal accountability relationships with their various stakeholders. For example, organizations engage in relationships with their professionals, managers and employees, contracting or trading working conditions, wages, and benefits for productivity and managerial judgment, where each party is accountable to the other for these expectations. Organizations engage in similar relationships with clients or customers, providing health care and other services, often for remuneration, with Boards of Directors to whom the organization is accountable, and in for-profit organizations, with shareholders, trading growth and profitability for capital. But like all organizational intentional behavior, these relationships are between collections of individuals acting on behalf of the organization and the respective stakeholders.

The result is that organizations exhibit intentional behavior, engage in reciprocal accountability relationships, are subjects of certain legal rights, and are said to act. But their so-called intentions, their accountability relationships, and their "actions" are the collective result of decisions made by individual persons. Their rights are assigned to an artificial entity, not to any individual person. The organization is an *eliminatable* subject because without persons, these "actions" literally could not occur. Thus, organizations are moral agents, but not moral persons.

Except for very small organizations, decision-making and action are often a result of the functions of disparate groups within the organization, e.g., physicians, nurses, managers, and administrators, so that an action of one part of an organization may be a function in part, of actions of another group within an organization. Often the final set of action plans or policies, evolving from combinations and permutations of individual primary actions done on the behalf of professional or impersonal organizational aims, cannot be identified merely with the sum of the original individual tasks. This is because of the role-related anonymity, uncertainty, and bounded rationality of the individual actions. In addition these individual actions can be changed through organizational culture, climate, and actions of other individuals and groups in other divisions of the organization, and the ways in which goals are interpreted at each stage of activity. Thus the *reasons* for organizational "actions" often cannot be explained merely by itemizing individual constituent actions. The result is organizational "activity" or policy that is a secondary action, an action ascribed to the collective, the organization, even though the collective itself did not literally authorize the action. Thus collective action may not be identical to individual action because of the role-related nature of professional, clinical, and managerial decision-making and the impersonal character of the authorization process.

Organizations, like individuals, do not always "respond" positively or negatively to moral pressures, because organizational moral agency is not independent of the moral input of its culture, and its professional, administrative, clinical, and managerial staff. Moral reactions of persons are necessary (but not sufficient) for collective moral reaction. The kind and degree of organizational moral "action" and moral "response" obviously depend on the kinds and degrees of primary constituent moral actions and reactions. For example, many employees at the University of Virginia Medical Center did not feel responsible in any way for the baby swap. Therefore, the Medical Center as a collective did not initially "respond" morally to allegations about the switch. Because they were unable to trace the origin of actions that resulted in the baby switch, the University originally contended that the switch could not have happened by accident, and implied that it was a deliberate criminal act (Blum and Shear, 1998, p. A-4). This explanation does not excuse the University from moral culpability, but this pathology helps to explain how an organizational culture and role assignments can create moral blindness. Later, of course, the University recognized its culpability even though it could not trace the alleged criminal culprits, apologizing to the families and children who were affected (Shear, 1998, p. B-1). Still, although they sometimes appear to take morally neutral stances, and do not always "recognize" moral demands, organizations as collectives are made up of persons who can "act" on behalf of the organization, thus creating its moral agency, and thus its moral responsibility. Moral blindness does not excuse an organization from moral responsibility just as it does not excuse rational free individual moral agents.

2.2 The Profit Motive and Friedman Economics

Much of health care is managed through HMOs, many of which are for-profit organizations. Thus it is tempting to appeal to concepts in business ethics to sort out the complex issues that arise in these settings. One fear of using tools from business ethics to analyze organizational and systemic issues in health care, however, is that the profit motive, allegedly in the forefront of management thinking in for-profit corporations, will emerge as the allegedly dominating motive in health care organizations, a motive that may undermine the successful financing and delivery of health care. However, I want to argue that (a) the profit motive is not even a dominant motivator in the best companies and (b) a stakeholder approach will further belie this perception while at the same time creating a viable model for dealing with healthcare organizations (HCOs).

Milton Friedman some time ago declared:

> There is one and only one social responsibility of business—to use its resources and engage
> in activities designed to increase its profits so long as it stays within the rules of the game,
> which is to say, engages in open and free competition without deception or fraud.
> (Friedman, 1970, p. 126)

This often-misquoted statement does not advocate that "anything goes" in commerce. Law and common morality should guide our action in the marketplace just as they guide our actions elsewhere. Nevertheless, given that qualification, which is an important one, managers' first duties and fiduciary duties are to owners or shareholders. Ordinarily these duties are to maximize return on investment, although in some companies the mission statement directs managers to other ends as well.

Friedman's depiction of the relationship between ethics and business has been influential in changing the model of contemporary healthcare delivery. The promise of managed care has been that self-interested commercial competition between providers and insurers will be a sufficient mechanism to improve the efficiency and reduce the cost of health care, without imperiling quality. Yet there are a number of difficulties with this argument even as it applies to the practice of commerce or business, and even greater difficulties when applied without qualification to healthcare management and delivery.

One difficulty is that, in fact, many of the best for-profit corporations do not operate under Friedman's philosophy. In a six-year project, James Collins and Jerry Porras, set out to identify and systematically research the historical development of a set of what they called "visionary companies", to examine how these companies differed from a carefully selected control set of comparison companies (Collins and Porras, 1994, p. 2). Collins and Porras defined the visionary company as the premier organization in their industries, as being widely admired by their peers, and as having a long track record of making a significant impact on the world around them (p. 3).[5] What was different about visionary companies and comparison companies? Each operates in the same market and each has relatively the same opportunities. Still, Collins and Porras, state:

> Contrary to business school doctrine, "maximizing shareholder wealth" or profit maximization" has not been the dominant driving force or primary objective through the history of the visionary companies. Visionary companies pursue a cluster of objectives, of which making money is only one – and not necessarily the primary one. Yes, they seek profits, but they are equally guided by a core ideology – core values and a sense of purpose beyond just making money. Yet, paradoxically, the visionary companies make more money than the more purely profit-driven comparison companies. (p. 8)

Having dispelled Friedman's edict as the only acceptable normative framework for business organizations, there *is* one sense in which Milton Friedman's version might be useful in thinking about healthcare organizations. Healthcare organizations are, at least in theory, created for one purpose: to manage, evaluate, finance, and/or deliver health care to patients and patient populations. If their mission is patient or population health, then as rational agents they should act so as to maximize the treatment and well being of their designated populations. Rewording Friedman,

> There is one and only one social responsibility of any healthcare organization: to use its professional and economic resources and engage in activities designed to treat and improve the health of its patient populations so long as it stays within the rules of the game.... (Werhane, 2000, p. 173)

This formulation puts in perspective the unique feature of healthcare organizations that distinguishes them from other types or organizations including for-profit non-health related corporations. Actions of a provider organization that do not maximize patient or population treatment, or actions of an HMO that do not manage the delivery of that treatment effectively and successfully over a defined population, would, on this account, be irrational and indeed, morally wrong, given the mission of these healthcare organizations (Werhane, 2000).

Still, one of the "rules of the game" in the present economic climate might be the proviso that these organizations must be economically viable, that is, minimally, they must break even or create the ability to pay their debts. Even charity hospitals are under such economic constraints. Thus, even given a reformulated edict of Milton Friedman, can we simply subsume all healthcare organizations including provider organizations and HMOs, under the philosophical umbrella of commerce? Several characteristics of *healthcare* organizations complicate or even preclude such a move. Stakeholder theory helps us to sort out these complications, and an analysis of healthcare markets helps to distinguish health care from other types of market-driven organizations.

2.3 Stakeholder Theory and Stakeholder Priorities

A stakeholder is any individual or group whose role-relationships with an organization
  (a) helps to define the organization, its mission, purpose, or its goals, and/or
  (b) "is vital to the survival, and success [or well-being] of the corporation." (Freeman, 1999, p. 250), or
  (c) is most affected by the organization and its activities.

Let us assume for our purposes that all stakeholders in question are individuals or groups (including institutions) made up of individuals. If stakeholder interests have intrinsic value, then, according to R. E. Freeman, the "father" of stakeholder theory, in every stakeholder relationship, the "stakes [that is, what is expected and due to each party] of each are reciprocal, [although not identical], since each can affect the other in terms of harms and benefits as well as rights and duties" (Freeman 1999, p. 250). Therefore stakeholder relationships are normative reciprocal relationships for which each party is accountable. Figure One illustrates some of those kinds of relationships in a typical healthcare organization.

One of the challenges of stakeholder theory is to evaluate and prioritize various stakeholder claims with each other and with the profitability (or economic survivability) criterion Friedman and other economists advocate. In healthcare organizations, however, because of their unique features, the prioritization of stakeholder and claims is relatively more clear-cut.

By the fact of being a *healthcare* organization, the primary stakeholders in any healthcare organization are, or should be, its patients or patient population it serves. The difference between garden-variety corporations and any healthcare organization (whether a for-profit organization or not), even HMOs, is the primary mission is always the financing, management, and/or delivery of health services to individuals and populations. Moreover, since healthcare deals with individual persons, patient autonomy and informed consent are part of this equation. Indeed, Dr. Post, the ER physician in the Worthy case, in his comments on the aftermath of the case, stated, "I acceded in Mr. Worthy's decision to decline my recommendation to have a CT scan…. I believe strongly in patient autonomy and the right to choose" (Davis, Post, Rogers et al., 1998, pp. S4-5; Ozar, et. al., 2001).

The basic value-creating activities of healthcare organizations are the management, payment, and delivery of patient care. Such delivery requires professional expertise, without which the organization will fail, whatever its goals. Therefore, healthcare professionals are the second most important stakeholders both for HMOs and provider organizations. Typically, healthcare professionals belong to, and are accredited by, independent professional associations. Many if not all professionals consider themselves primarily bound by the ethical prescriptions of their profession, preeminent among which are their duties to their patients. One of Dr. Davis's dilemmas in the Worthy case was the conflict between her professional duty to patient health and the capitation requirements of GoodCare, which affected her compensation and the compensation of her medical group. As she stated, in commenting on the case, "...[I]t's increasingly important (for my financial well-being and the plans') that I authorize only 'appropriate' referrals" (Davis, Post, Rogers et al., 1998, p. S3).

Financial stability is the third priority in healthcare organizations, which in for-profit organizations is translated into shareholder value, as shareholders are the third most important stakeholders. Long-term organizational viability that includes financial stability is necessary for the continuation of any of these organizations and part of the guarantee of the quality of its services, even charity organizations. Indeed, it is not odd that healthcare organizations are concerned with efficiency, profitability, or at least, economic survival. But the trouble begins when a healthcare organization realigns its mission or creates an organizational culture in which efficiency, productivity, and/or profitability become the overriding priorities, as may have been so in the Worthy situation. Dr. Depp's statement after the death of Worthy, "The HMO has also adhered to a firm policy of requiring primary care provider approval for all specialty referrals...to protect our members from getting unneeded specialty services" (Davis, Post, Rogers et al., 1998, p. S8) illustrates this preoccupation for rules and efficiency.

Part of professional responsibility as defined by every healthcare professional's code of ethics (and spelled out in the Code of the American College of Health Care Executives as well!) is commitment to community and public health. Despite the ability of HMOs to define and restrict the patient population they will serve, community access and public health are always part of the accountability equation, because of the simple societal expectation that healthcare organizations and professionals *should* serve public health needs. One of Dr. Post's problems as an ER physician was that his hospital was in a poor community. The commitment to that community forced him to consider whether to encourage Worthy to have a CT scan if it would not be reimbursed.

Stakeholder prioritization creates values prioritization as well in health care organizations. If the primary stakeholder is the patient and patient populations, then the first goal of any healthcare organization should be the autonomy and care of its patients. The second goal is the encouragement, respect, and well being of its professional staff and their decisions; the third is financial viability; and the fourth, but not least, community health.[6]

84

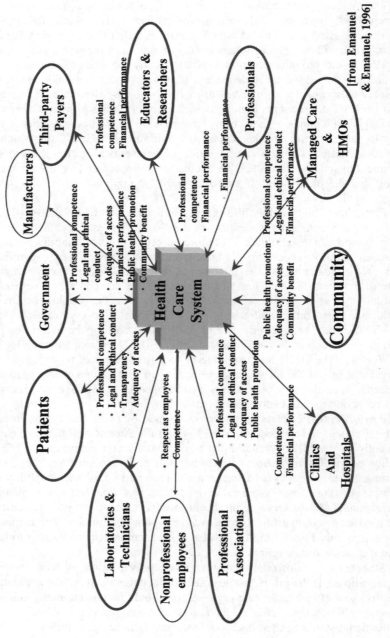

Figure 1: Stakeholder Relationships in our health care system

[from Emanuel & Emanuel, 1996]

Given these priorities as provisos, can we nevertheless treat healthcare organizations like other businesses? There is a further complication, the nature of healthcare markets.

2.4. Healthcare Markets

There are a number of factors that complicate healthcare markets, adding to the questionability of treating healthcare organizations univocally like other business corporations. These factors include the distinction between customers/payers and consumers/patients, information asymmetries, and the distortion of supply-and-demand models.

The correlation between consumers and payers is very different in this organization than in the usual business. In provider organizations, recipients of healthcare services are usually not the payers. Various forms of insurance, employer-sponsored health plans, or government agencies (the "customers") purchase health coverage for the individuals and patient groups who are the actual and potential patients (the "consumers"). This three-way relationship complicates accountability between the parties affected in healthcare delivery. Moreover, unlike the typical consumer, the patient, like John Worthy, may have limited choices in delivery options. "Buyer Beware" is not an appropriate slogan for healthcare consumers. Even in those cases where the patient is also the payer, the consumer/patient is often ill and vulnerable. So, unlike ordinary consumers, patients are not always able to exercise their choices coherently.

Coupled with patient vulnerability, healthcare patients are seldom "fully informed". There is an obvious information asymmetry between HMOs and provider organizations, between managers and healthcare professionals, and between professionals and customers or patients regarding knowledge about health. There is also an information asymmetry between healthcare organizations. Competitive HMOs do not have access to consumer (i.e., patient) information in ways in which they have access to market information in other business enterprises. So ordinary competitive relationships are not possible in the healthcare market and ordinary competitive information is not always available in the healthcare "market".

Finally, supply-and-demand curves are skewed in the healthcare industry. Healthcare organizations cannot respond to all market demands, in particular, to the demands of the uninsured. Some patients or patient-groups cannot pay for what they consume while others pay for more than they consume. These factors, and there are others, give ample evidence that the distinguishing features of healthcare organizations warrant their separate study. Business ethics provides some tools for that study, but this does not merit merely conflating HCOs with other commercial organizations.

Returning to the John Worthy case, does an organizational analysis of this situation help in sorting out this situation? Focusing on Worthy, his autonomy, his implicit consent to the healthcare plan, and his refusal to heed the recommendation of Dr. Post, the ER physician, because the hospital was not affiliated with GoodCare contributed to his death. The prioritization of patient health by Dr. Davis, GoodCare, and the emergency room, would have helped to alleviate this problem. But each of these people, and GoodCare itself, was functioning in a complex network or system of interrelationships all of which complicated the situation. Each individual was caught in the maelstrom of this system and no one could or did sort through and prioritize these relationships in ways that might have saved Worthy. GoodCare itself was mired in rules that helped most of its patients, but was ill-suited to emergency situations. Thus the case calls for a more complex analysis.

3. SYSTEMS AND SYSTEMS ETHICS

A truly systemic view [of current health care] …considers how [this set of individuals, institutions, and processes] operates in a system with certain characteristics. The system involves interactions extending over time, a complex set of interrelated decision points, an array of [individual, institutional, and governmental] actors with conflicting interests…and a number of feedback loops. (Wolf, 1999, p. 1675)

What do we mean by a *healthcare system?* There are at least two ways to think about systems that are helpful for our purposes. First, a healthcare system can be defined as "assemblages of interactions within an organization or between organizations" (Emanuel, 2000, p. 152). These "assemblages" are governed by (1) organization arrangements or structure, (2) attributes that organize and shape an institution, (3) the interrelationships between organizations, and/or (4) procedures and processes that are adapted by an organization or a set of organizations. Under this rubric healthcare systems are networks of relationships between individuals, between individuals and organizations, between organizations, and between individuals, organizations, institutions, and government. On a macro level a healthcare system, in the case of the United States, includes government agencies that regulate the healthcare industry and create public policy, government payers including Medicare and Medicaid, public policy (and public policy think tanks), insurers, employers and employer health plans, HMOs, MCOs, health care centers (including academic medical centers), hospitals, nursing homes, laboratories, pharmaceutical companies and pharmacists, healthcare networks, professionals and other care givers (both employed and independent professionals), researchers, writers, lawyers, healthcare executives, managers and other employees, professional and academic associations, patients and patient populations, and social workers. The system is defined by the myriad of networks and interrelationships between these institutions, organizations, employers, managers, professionals, caregivers, payers, patients, and managers. Thus systems under this rubric are social networks of interactions between individuals and groups of individuals within and across institutions.

A second approach to thinking about healthcare systems is best described by Laszlo and Krippner's definition of a system as "a complex of interacting components together with the relationships among them that permit the identification of a boundary-maintaining entity or process" (Laszlo and Krippner, 1988, p. 51). A "system" under this definition refers to a model or framework, a structured set of methodologies, rules or processes that frame a certain approach, protocol, or decision process. On this account, a systems approach embeds a model that includes a set of techniques or procedures to deal with a bounded class of phenomena. For example, processes to reduce house officer fatigue, to test blood, to diagnose certain diseases, or to catch medication errors are systems protocols that function similarly within and/or across organizations. In framing the process or prescribing the set of rules for procedure, this approach also limits or proscribes as well so that the set of rules are boundary-maintaining in Laszlo and Krippner's sense of the term. These processes or rules are protocols that are adapted by many individuals and organizations engaged in the procedures in question, and which protocols are adapted will obviously affect the network of individuals and institutions where they are adapted.

These two approaches to thinking about systems are interconnected. A systems protocol to catch medication errors might be adapted by a number of organizations and within each organization that adaptation helps to define and frame the way in which that institution operates and how its operations affect its professionals and patients. Rules, procedures, and other arrangements within a particular organization shape which protocols are adapted for medical procedures and how those are interpreted, and also affect or are affected by HMOs, MCOs, and other insurers or payers of healthcare. An HCO that has a reputation for repeated medication errors, for example, will likely attract attention to insurers and other payers of patient care at that organization and eventually affect that organization and its structure. Conversely, the structure and mission of a particular organization of a healthcare system may preclude the adoption of certain protocols. For example, capitation or coverage limitations by insurers or the government on treatment may preclude the practice of certain experimental protocols, or, in the case of Worthy, the use of an unspecified emergency room.

What is characteristic of both types of systems is that any phenomenon or set of phenomena that are defined as a system has properties or characteristics that are lost or at best, obscured, when the system is broken up into components. For example, in studying Worthy's healthcare system, merely analyzing GoodCare will obscure many of the issues in the case. Similarly, in viewing organizations such as GoodCare, if one focuses simply on its organizational structure, or merely on its mission statement, or only on its employees or participants, one loses the interconnections and interrelationships that characterize that system or subsystem. Similarly, if one isolates one component of a protocol for study, one is no longer studying the protocol (Laszlo and Krippner, 1998, p. 53). The kinds of systems we are concentrating on in this paper, healthcare systems, have another characteristic. Each type of system or subsystem is purposive or goal-oriented. Protocols and procedures are developed with outcomes in mind, and a good procedure is one that produces clear and consistent outcomes as defined by the structure and methodology of the procedure. Indeed, the prescribed outcomes of any protocol create the boundaries for that protocol. Organizations and institutions also have goals that are usually reflected in their mission statements or other statements of purpose. When one looks at the macro system of health care in this country, the goals are confusing and less clear. Nevertheless one usually ascribes goals to healthcare systems, e.g., patient, patient populations, and/or public health, although this ascription is sometimes a normative prescription rather than a description of the purposes and goals presently in place.

The goal-orientation of healthcare systems accounts for their normative dimensions. As has been argued extensively earlier, organizations as well as individuals have purposes and goals that carry with them moral obligations, and we hold organizations and institutions, as well as individuals, morally accountable (French, 1979; Werhane, 1985; Spencer, et. all, 2000; Ozar et al., 2001). While it is less transparent that systems, too, are moral agents of some sort, it is true that the structure, interrelationships, and goals of a particular system produce outcomes that have normative consequences. An alteration of a particular system or parts of that system will often produce different kinds of outcomes. A system is comprised of a set of networks of relationships between individuals; thus a system is developed from relationships between people. How the system is construed and how it operates affects and is affected by individuals. The character and operations of a particular system or set of systems affect and is affected by those of us who come in contact with the system, whether we are patients, the community, professionals, managers, or insurers. That we do not have universal healthcare coverage in the United States, for example, not only affects the uninsured, but also has economic and social consequences for communities, healthcare providers, and professionals. Finally, systems as well as organizations and individuals are often causally responsible for healthcare delivery and clinical performance. The systems of payment for healthcare delivery, for instance, affect the quality and kind of delivery available to many patients, as Worthy discovers. Thus moral responsibility is incurred by the nature and characteristics of the system in question (Emanuel, 2000).

Systemic arrangements and organization networks create roles and role responsibilities, rights, and opportunities that affect individuals and individual activities and performance. What is less obvious is that one can take a single organization or a single individual functioning within that organization or system and apply different systems matrices to that organization with differing outcomes. The preoccupations of subsystems and individuals functioning within these systems and the ways values and stakeholders are prioritized affects the goals, procedures, and outcomes of the system or subsystem in question. On every level, the ways we frame the goals, the procedures and what networks we take into account makes a difference in what we discover and what we neglect. These framing mechanisms will turn out to be important normative influences of systems and systems thinking.

For example, a salaried healthcare professional, such as Dr. Davis, operating within a strict capitation HMO may see her role as a professional differently than, say, an individual physician in a small independent practice. Indeed, different mindsets operative in the Worthy case prioritize events and their importance differently, thus creating part of Worthy's problem. Sometimes, as in this instance, the professional model outlined by the Hippocratic Oath is replaced by or blurred with economic interests. Then individual patient need as determined by healthcare professionals is evaluated according to the criteria of efficiency and cost, and attention to individual patients is replaced by a group identification to a defined patient population and carefully prescribed rationing. The Worthys were concentrating on themselves and what they took to be an individual patient-physician relationship with Dr. Davis. Dr. Davis, with a bloated caseload and pressures to see as many patients as possible, focused on current patients, the exigencies of that day, and GoodCare's capitation requirements. Dr. Newman's office was preoccupied with the HMO rules, while the emergency room was concerned with Worthy's coverage as well as his autonomy. Depp and GoodCare were interested in patient populations, costs and benefits of healthcare financing and delivery, and what they could deliver productively and with efficiency to their defined population. These mindsets and their priorities set differing boundary conditions that scarcely overlapped in ways that could have benefited Worthy. Different perspectives frame the goals and expectations of the HCO differently with obviously differing outcomes.

3.1 Systems Thinking

What do we mean by "systems thinking" or a "systems approach"? For our purposes, systems thinking presupposes that most of our reasoning, experiencing, practices, and institutions are interrelated and interconnected. Almost everything we can experience or think about is in a network of interrelationships such that each element of a particular set of interrelationships affects the other components of that set and the system itself, and almost no phenomenon can be studied in isolation from all relationships with at least some other phenomenon. Systems thinking, then, involves two kinds of analysis. In a systems approach, "concentration is on the analysis and design of the whole, as distinct from...the components or parts..." (Ramos, 1969, pp. 11-12). Systems thinking requires conceiving of the system as a whole with interdependent elements, subsystems, and networks of relationships and patterns of interaction. Studying a particular component of a system or a particular relationship is valuable under this rubric only if one recognizes that that study is, at least in part, an abstraction from a more systemic consideration.

Secondly, few systems are merely linear and few, including healthcare systems, are closed systems that are not constantly in dynamic processes of changing and reinventing themselves. Because "the fundamental notion of interconnectedness or nonseparability forms the basis of what has come to be known as the Systems Approach,...every problem humans face is complicated [and] must be perceived as such" (Mitroff and Linstone, 1993, p. 95). So each system or subsystem, because it is complex and entails a multitude of various individual, empirical, social, and political relationships, needs to be analyzed, and analyzed from multiple perspectives. The downside of not taking this sort of approach is obvious in the Worthy case. Looking at Worthy's situation merely from his perspective and that of his wife was disabling, just as disabling as examining this problem merely from the perspective of Dr. Davis or Dr. Post.

A Multiple Perspective method postulates that any phenomenon, organization, or system (or problems arising for or within that phenomenon or system) should be dealt with from a number of perspectives, each of which involves different world views where each challenges the others in dynamic exchanges of questions and ideas. Mitroff and Linstone suggest that in business, economic, and public policy contexts one needs to look at problems from a technical, or fact-finding point of view, from an organizational or social relationships perspective, and from an individual perspective, ranking problems, perspectives and alternate solutions, and evaluating the problem and its possible resolution from these multiple perspectives (Mitroff and Linstone, 1993, chapter 6). While it is never possible to take into account all the networks of relationships involved in a particular system, and surely never so given that these systems interact over time, a multiple perspectives approach forces us to think more broadly, and to look at particular systems or problems from different points of view. This is crucial in trying to avoid problems such as Worthy's because each perspective usually "reveals insights...that are not obtainable in principle from others" (Mitroff and Linstone, 1993, p. 98). It is also invaluable in trying to understand other points of view. A Multiple Perspectives approach is essential if, for example, for-profit healthcare systems are to understand what is at stake for the uninsured, or what is at risk if professional staff is overburdened with efficiency requirements or is reduced. This kind of approach is critical for not-for-profit healthcare organizations to help them understand the economic dynamics of healthcare delivery, and such an approach is important for healthcare consumers so that people like John Worthy can comprehend what sort of system they have enrolled in.

3.2 Multiple Perspectives Approach to Healthcare Systems

In examining ethical issues in healthcare systems, subsystems, and organizations, a Multiple Perspectives approach might involve describing the system and subsystems in question, and an examination and evaluation of the network of organizational, relational, and individual perspectives. One investigates how a particular configuration of a system or subsystem affects individuals, in this instance, healthcare delivery, community and public health, and micro issues such as patient autonomy, access, informed consent, privacy, and other important matters of concern to those delivering and receiving care. One evaluates the boundary creating processes, to be clear about what is left out, as well as accountability relationships, as we illustrated earlier. Then one evaluates both by prioritizing the goals of the system and indeed, by evaluating those goals (including the professional, organizational, economic and sometimes, political norms) implicit in the system.[7] The aim is three-fold: to understand the system and its complex interrelationships; to evaluate that system, its relationships, and the moral responsibilities embedded therein; and to think through possible solutions or resolutions of the issues in question that take into account these multiple perspectives and individual as well as organizational and systems responsibility.

The first, a descriptive or "technical" perspective, includes the following. First one describes the system in question, as we tried to outline the U.S. healthcare system in the beginning of the paper, and the health care system in which Worthy was involved. Included in the description are networks of interrelationships between individuals, groups, organizations, and systems, and the number, nature, and scope of subsystems in the system in question. Stakeholder analysis is useful in this context. By enumerating the various stakeholders involved in or affected by this system, and their interrelationships and accountabilities, one can get clearer on the complex nature of the system and its outcomes. Worthy's employer presented him three healthcare alternatives. He accepted one alternative, mainly for cost reasons, without considering the positive aspects of other choices. His primary physician, Dr. Davis, was part of that GoodCare system, about which she knew very little. She was measured by GoodCare on volume of patients seen and on capitation criteria that restricted her decision-making. GoodCare itself had developed a series of rules and regulations that prevented it from being clear on its alleged mission to health. Dr. Post, the ER physician, made choices on the basis of patient autonomy but his choices were also affected by the fact that his hospital took in many indigent and working poor patients and needed to be careful to seek out paying patients as well. So the economic pressures on these stakeholders affected their decisions, even Worthy's.

In this process it is also helpful to outline the boundaries and boundary-creating activities so that it is clear what is not included in the system. But different stakeholders will outline the boundary conditions differently because of the way they prioritize the value creating activities of the system. For example, Fran Davis, the Worthy family doctor, said, "I made the decision not to see Mr. Worthy in the office and recommended instead that he go to the emergency room if the pain 'seemed that severe' … It's increasingly important (for my financial well-being and the plans') that I authorize only "'appropriate' referrals" (Davis, Post, Rogers et al., 1998, p. S3). In other words, she prioritized the patients on her schedule and her financial well-being ahead of what turned out to be a critical situation. Similarly, Dr. Post claimed, "I believe strongly in patient autonomy and the right to choose…" and argues further, "the chance that those tests [Worthy's] would not be reimbursed poses a dilemma…." (p. S5). So Dr. Post prioritized patient autonomy and community health provision ahead of Worthy's particular condition. Connie Rogers, the customer service representative at GoodCare, who would not refer Mrs. Worthy to the President of GoodCare, had other priorities. One was the GoodCare rules for referrals. The second was the company president's edict that "he wants us to offer him solutions, not problems" (p. S6). Note that both physicians and Ms. Rogers were each trying to do his or her best in the situation. But this illustrates how prioritization of stakeholders and values (including economic values) changes the perception of boundary conditions and affects decision-making.

Linked to the boundary conditions and stakeholder prioritization are the accountability relationships between each stakeholder and element of the system in question. It is tempting to conceive those dyadically as Figure 1 illustrates, and from an organizational approach a dyadic description of accountability may be adequate.

But healthcare organizations are parts of more complex systems, and these relationships are much more overlapping and interlocking. See Figure Two for a partial diagramic depiction of some of these. Being clear about these relationships, and how each individual and each element of the system are or should be accountable to each other helps to clarify where decisions go wrong.

Imbedded in this process are the goals or purposes of the system, or in the case of health care or a healthcare subsystem, what goals it *should* have, and how these are prioritized, since the goals a system has will affect its structure, interrelationships, and outcomes. These prioritized goals then become the evaluative elements overlaid on the descriptive grid. In the Worthy case, the goal of any part of this system should be first, and primarily, patient well-being. Yet these were sometimes confused with rule-following and efficiency at GoodCare. The reward system for its physicians was based on a capitation system that seemed to allow no loopholes for emergencies. Dr. Post's hospital was in financial trouble, so his worries included those economic issues as well. GoodCare is not an evil organization bent on eliminating its consumer-patients. Dr. Davis and Dr. Post are caring, well-trained, physicians. Economic viability and financial survival are not wrong-headed aims. Following rules is ordinarily a good way to operate. However, it is the ways these elements are prioritized that gets GoodCare and its physicians in trouble. Yet if this system had an agreed upon overlapping set of goals, if each was clear on what these were and how they were to be prioritized, then patient well-being and professional judgment might have been placed ahead of rules and economic and efficiency concerns.

There is one more perspective to consider in the Worthy case, that of individual responsibility, the responsibilities of the professionals, managers, and of John Worthy himself. A systems approach should not be confused with some form of abdication of individual responsibility. As individuals we are not merely the sum of, or identified with, these relationships and roles; we can evaluate and change our relationships, roles, and role obligations, and we are thus responsible for them. That is, each of us is at once a byproduct of, a character in, and the author of, our own experiences. The physicians and other healthcare professionals in the Worthy scenario each had responsibilities as defined by their profession, the responsibility to patient care and well-being. While employment demands may have conflicted with that responsibility, as professionals, the Hippocratic Oath is supposed to take precedence. Interestingly too, many healthcare executives now are members of a professional association as well, the American College of Healthcare Executives. Part of that professional code states, "[t]he fundamental objectives of the healthcare management profession are to enhance overall quality of life, dignity, and well being of every individual needing healthcare services" (American College of Health Care Executives, 2001). The GoodCare organization, at least as evidenced in this case, created barriers to these professional responsibilities. Still the professional "actors" in this case, by accepting their roles as professionals, had responsibilities that should have taken precedence.

Figure 2: Stakeholder Systems Networks

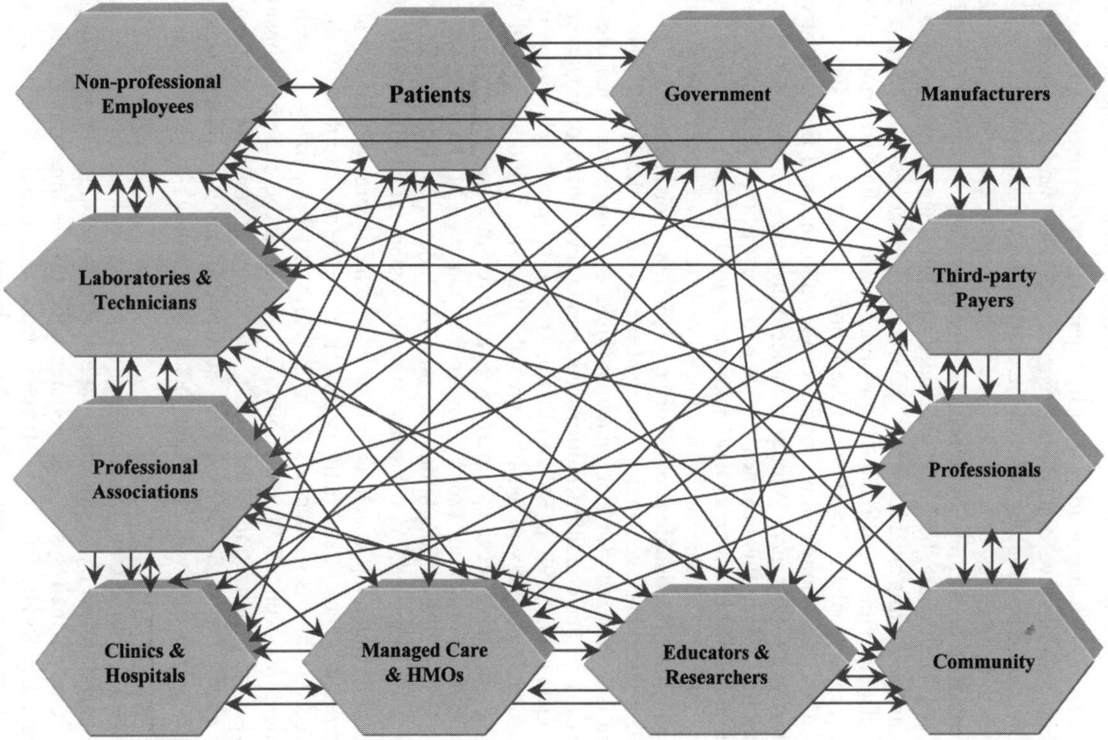

John Worthy, too, had responsibilities—responsibilities to be familiar with his health coverage and its limits, and to make an appointment to see Dr. Davis before he became ill. He, too, was responsible, and we cannot mitigate all of that responsibility through blaming the health system in which he found himself.

We can now evaluate John Worthy's particular dilemma and the problems at GoodCare. Earlier we argued that part of the issue was a misprioritization of stakeholders and stakeholder concerns in creating organizational goals at GoodCare, with the capitation pressures on Dr. Davis, the hospital economic pressures on Dr. Post, and cost constraints of Worthy's employer, who simply "farmed out" its employees' health care, because "we…need to control our operating costs, of which health care premiums are a large part" (Davis, Post, Rogers et al., 1998, p. S10) But simply placing patient and patient population health as the first priority does not in itself solve the problem. Everyone in the Worthy scenario thought they *were* putting patients first, even though some confused rule-following and financial survival with that priority. Dr. Post gave Worthy a choice, but given his condition, it was not an informed one. Given this healthcare system, the first priority of patient health has to be operationalized in a manner that makes sense to all parties under the pressures the ordinary patient, healthcare professional, or executive faces every day. One approach is the "reasonable person" or "prudent layperson" standard. This standard, in brief, argues that even under managed care, one of the "rules" should be that payment is authorized "for emergency and other services when a reasonable person believes his or her condition requires emergency evaluation or treatment" (Davis, Post, Rogers et al., 1998, p. S6). While this standard appears to be fuzzy, it turns out that reasonable people will usually agree about such conditions. We all agree that Worthy should have received emergency care even in a unit not ordinarily covered by GoodCare. But neither Worthy nor Dr. Post nor even Michael Depp, the Medical Director at GoodCare, thought this was possible. Indeed, Dr. Depp argued, "Looking back on what happened, I don't really think our system failed Mr. Worthy" (p. S9).

Another related approach is to consider which core values are at stake in healthcare management and delivery, prioritize those, and use those as the basis for one's decision in crisis situations such as Worthy's. Prioritizing the well being of the patient (as well as his autonomy) is and must be the first considerations. Otherwise the organization will fail as a *healthcare* organization. Moreover, if the Collins and Porras study has validity, such prioritization of patient wellbeing and professional expertise will allow healthcare organizations to succeed, even become profitable.

But let us assume that GoodCare and its professional and managerial staff know all of this and in fact try to prioritize patient care and respect for professionals, even though some of their systems procedures belie this assumption. There is a third challenge. Dr. Davis, Dr. Post, Michael Depp, and Connie Rogers are all in the middle of large organizations—they are employees. How do they lead "from the middle" and make what we, after the fact, declare as the "right" decisions? If, as professionals they have duties specified by their professions, and if these are to override other claims, then, first, they have to be operationally committed to that prioritization.   Secondly, as professionals, they have to look for allies within GoodCare, challenge the rule-bound status quo, and make a space for the value of patient care. While this appears to be asking a great deal of these individuals, in fact there is evidence that such forms of leadership are very effective in challenging organizational authority and systems constraints (Bruner, et. al., 1998, pp. 251-275; Nielsen, 2000). Such leadership was patently absent in John Worthy's situation. If only one of these professionals had spoken out, raised the issue of patient priority, and taken professional responsibility, John Worthy might be with us today.

To summarize, what I am suggesting is that what was lacking in the Worthy case was self-awareness of the systems in which these actors were operating, evaluation of the system, organizations, and sub-systems in question, and then, finally, courage to make changes both in decision-making and later, to organizations in question and to the system itself. What is missing in this scenario is what I have called in another place, "moral imagination", "the ability in particular circumstances to discover and evaluate possibilities not merely determined by that circumstance, or limited by its operative mental models, or merely framed by a set of rules or rule-governed concerns" (Werhane, 1999, p. 93). In systems thinking what is required is moral imagination, now operating on organizational and systemic levels as well as within individual decision-making.

3.3 Systems Thinking and Traditional Health Care Issues

Let us step back from the Worthy case and consider another application of systems thinking. Given the present state of healthcare payment and delivery in the United States, a systems approach is essential in dealing adequately with more traditional ethical issues in medicine. In a recent article on informed consent Susan Wolf (1999) uses that example to argue that merely considering patient informed consent by the physician at the point of treatment (a dyadic approach) is an oversimplification of the complex arrangements in health care systems and bypasses many elements that affect patient consent.   Systemic arrangements demand informed consent concerning subscription to a plan, insurance and healthcare coverage, patient information sharing and transfer, rationing, and other points of care. It is precisely in this network or web of relationships that patients become ill-informed about their coverage, options, insurance, etc. Merely to think of informed consent in the patient-healthcare professional one-on-one encounters will not take into account how that patient is insured, how the professional and the healthcare center are reimbursed, capitation and rationing limitations (which may be proscribed by the employer of the patient, the insurer, the state, Medicare or Medicaid, or by other criteria), patient options in treatment or choices to go to another healthcare center. Obligations to inform patients run all the way through the healthcare system and not merely at entry or before treatment. It is not merely the *physician's* duty to inform her patients since that is placing an undue burden of knowledge on healthcare professionals. These duties run throughout all parts of the system as well. Disclosure and full information needs to be disseminated at every stage, not merely at the point of treatment (Wolf, 1999).

A similar analysis reveals the complexity of protecting the privacy of medical records. In analyzing privacy issues, traditionally the focus was often on the protection of patient records by the physician or the hospital. Under the present health care delivery system, this, in fact, has little effect on protecting these records. Indeed, even in the "olden days", one's records were subject to review by those working in the physician's office, by nurses, interns, residents and other attendants at the hospital, one's pharmacy, laboratory, and by any other medical facility with which one had contact.[8] Today, one's insurance company, employer, laboratory, hospital or other healthcare center usually has access to these records. Insurance databases often have access to information. Records of those on Medicaid and Medicare or claiming disability or workers compensation are available to state and federal government agencies. Anyone using the internet for health advice or for ordering prescriptions exposes herself to information sharing. While federal medical records safeguards are being proposed, these would require patient written consent before disclosure in even the most routine cases such as claims payment. Such legislation might protect patient privacy to some extent, but operationalizing written consent in practice might create overwhelming complications. And since every patient would have to give consent for her treatment and insurance, the result would not be to protect the privacy of medical records except in some instances from managed care organizations other than one's own, or perhaps from online advertisers and marketers who often share consumer/patient lists. What appears to be simple protection of a patient's records in fact must be approached systemically in order to comprehend what is at stake and to achieve any results (Murray, 1999, p. A-12; Pear, 2000, p. A-1; Hodge, Gostin, and Jacobson, 1999).

To conclude, systems thinking is essential if we are to understand, evaluate, and change healthcare payment and delivery in this country. Nevertheless, despite the importance of systems thinking and systems analysis in health care, and this is the final point of the paper, no healthcare system or subsystem is or need be thought of as a closed static system. Healthcare systems are dynamic and revisable phenomena, created and changed by individuals. But until we comprehend the complexity of the systemic interrelationships within and across systems, we cannot successfully evaluate the system or subsystem in question and begin to make changes that are critical if we are to avoid unnecessary deaths such as John Worthy's. Nor, if Susan Wolf is right, can we deal adequately with traditional ethical issues in health care. This paper is an initiation of that set of thinking processes.

University of Virginia
Charlottesville, Virginia, USA

NOTES

[1] This paper has benefited greatly from the work of Linda Emanuel and Susan Wolf, and from collaborative work with David Ozar, Ann Mills, Mary Rorty, and Edward Spencer, and the comments of Norman Bowie and Andy Wicks. An earlier version of this paper appears in *The Blackwell Guide to*

Business Ethics, edited by Norman E. Bowie (Malden, MA: Blackwell Publishers, 2002, pp. 289-312). Reprinted by permission of the editor and the publishers.

[2] Reprinted by permission from the *Hastings Center Report, Special Supplement*, July-August 1998, pp. S1-S3.

[3] This section is revised from Freeman and Werhane (2002). Reprinted by permission of the authors, editor, and publisher. The original work for this section appeared in Werhane (1985). Reprinted by permission of the author, Patricia Werhane.

[4] It is tempting to make a mistake here and claim that because corporations can be moral agents they can be moral persons. Werhane (1985) contains the arguments as to why this is a mistaken analogy.

[5] The long-term financial performance of each has been remarkable. A dollar invested in a visionary company stock fund on January 1, 1926, with dividends reinvested, and making appropriate adjustments for when the companies became available on the stock market, would have grown by December 31, 1990 to $6,356. A dollar invested in a comparison stock fund composed of these companies would have returned $955 – more than twice the general market but less then one sixth of the return provided by the visionary companies. That dollar invested in a general market fund would have grown to $415 (Collins & Porras, 1994, pp. 4-5).

[6] Ozar, et. al. (2001) postulate other priorities including unmet healthcare needs, advocacy for social policy reform and community benefit. I have gathered those together in the "community and public health" category. Ozar, et. al. also place organizational solvency and survival in the list of secondary priorities. I disagree with this prioritization.

[7] In analyzing the ethics of systems, Linda Emanuel proposes an evaluative grid that sets out the purpose, structure, processes and outcomes of a particular system against professional, political, and economic models (Emanuel, 2000, Table 1, p. 161). While the details of that grid are certainly subject to more debate, this approach pushes us into the direction of more broad-based systems thinking and into more creative and imaginative ways to analyze and evaluate healthcare systems. In the case of Worthy, we are working on a more micro level, dealing with issues within the United States healthcare system and the system of GoodCare. Moreover, using Emanuel's grid, one has to sort out and evaluate professional, political, and economic models as they apply to health care. Our prioritization in the previous section of patient health and autonomy, professional competence and excellence, economic viability, and public health, is one example of such an evaluative scheme. While many will argue as to the sequencing of these priorities, some sort of prioritization is implicit in any list of professional, political, and economic norms, and making those explicit helps to clarify where there is congruence.

[8] A physician writing about confidentiality of patient information did an informal survey and found that 75 clinicians or employees had legitimate access to his patient's record, which meant that they were in some measure engaged in his patient's care (Siegler, 1982).

BIBLIOGRAPHY

American College of Healthcare Executives (2001). 'Professional code.' [On-line] Available: www.ache.org

Bayley, C. (2001). 'Turning the Titanic: Changing the way we handle mistakes.' *HEC Forum, 13*, 148-159.

Berwick, D.M & Nolan, T.W. (1998). 'Physicians as leaders in improving health care.' *Annals of Internal Medicine, 128*, 289-292.

Blum, J. & Shear, M.D. (1998 August 4). 'Family vows not to uproot swapped girls.' *Washington Post*, A-4.

Bowie, N. (ed.) (2002). *Blackwell's Guide to Business Ethics*. Malden, MA: Blackwell Publishers.

Bruner, R., Eaker, M., Freeman, R.E., Spekman, R. & Teisberg, E.O. (1998). 'Leading from the middle.' *The Portable MBA* (251-275). New York: Wiley.

Collins, J. & Porras, J. (1994). *Built to Last*. New York: HarperBusiness.

Emanuel, L. (2000). 'Ethics and the structures of health care.' *Cambridge Quarterly of Healthcare Ethics. 9*, 151-168.

Emanuel, E. & Emanuel, L. (1996). 'What is accountability in health care?' *Annals of Internal Medicine*, 124, 229-239.

Freeman, R.E. (1999). 'Stakeholder theory and the modern corporation.' Reprinted in: Donaldson, T. & Werhane, P. (eds.), *Ethical Issues in Business, 6th edition* (pp. 247-257). Upper Saddle River, NJ: Prentice-Hall, Inc.

Freeman, R.. Edward and Werhane, P.H. (2002). 'Corporate responsibility,' In: Frey, R. and Wellman, C. (eds.) *A companion to applied ethics*. Malden, MA: Blackwell.

French, P. (1979). 'The corporation as a moral person.' *American Philosophical Quarterly, 16*, 207-215.

Frey, R. & Wellman, C. (2002). *A companion to applied ethics*. Malden, MA: Blackwell Publishers.

Friedman, M. (1970 September 13). 'The social responsibility of business is to increase its profits.' *New York Times Magazine*, 122-126.

Gibson, J.E. (1991). *How to do systems analysis*. Unpublished.

Hall, R.T. (2000). *An introduction to healthcare organizational ethics*. New York: Oxford University Press.

Davis, F., Post, E., Rogers, C. et al. (1998). 'What could have saved John Worthy?' *Hasting Center Report, Special Supplement, 28*(4), S1-S17.

Hodge, J.G., Jr., Gostin, L.O., Jacobson, P.D. (1999). 'Legal issues concerning electronic health information: privacy, quality, and liability.' *Journal of the American Medical Association, 282*, 1466-1471.

Inglehart, J.K. (1994). 'Physicians and the growth of managed care.' *New England Journal of Medicine, 33*, 1167.

Laszlo, A. & Krippner, S. (1998). 'Systems theories: Their origins, foundations and development.' In: Jordan, J.S. (ed.), *Systems Theories and a Priori Aspects of Perception*. Amsterdam: Elsevier.

Mitroff, I.I. & Linstone, H. (1993). *The unbounded mind*. New York: Oxford University Press.

Murray, Shailagh. (1999). 'On medical-privacy issue, the doctor finally may be in.' *Wall Street Journal*, August *20*, A12.

Nielsen, R.P. (2000). 'The politics of long-term corruption reform.' *Business Ethics Quarterly, 10*, 305-317.

Ozar, D., Emanuel, L., Berg, J., Werhane, P. & AMA Working Group on the Ethics of Healthcare Organizations. (2001). 'Organization ethics in health care: Toward a model for ethical decision-making by provider organizations,' Chicago: American Medical Association.

Pear, R. (2000 December 20). 'Clinton will issue new privacy rules to shield patients.' *New York Times*.

Ramo, S. (1969). *Cure for chaos*. New York: D. Mackay Co.

Shear, M.D. (1998 August 14). 'Baby's mother zeroes in on time that switch could have occurred.' *Washington Post*, p. B 1.

Siegler, M. (1982). 'Confidentiality: A decrepit concept.' *New England Journal of Medicine, 307* (24). 1518-1521.

Spencer, E., Mills, A., Rorty, M. & Werhane, P. (2000). *The ethics of healthcare organizations*. New York: Oxford University Press.

Werhane, P.H. (1985). *Persons, rights, and corporations*. Englewood Cliffs, NJ: Prentice-Hall Inc.

Werhane, P. H. (1999). *Moral imagination and management decision making*. New York: Oxford University Press.

Werhane, P.H. (2000). 'Stakeholder theory and the ethics of healthcare organizations.' *Cambridge Quarterly of Healthcare Ethics, 9*, 169-181.

Wolf, S. (1999). 'Toward a systemic theory of informed consent in managed care.' *Houston Law Review, 35*, 1631-1681.

Wong, K.L. (1999). *Medicine and the Marketplace*. Notre Dame IN: Notre Dame University Press.

GERALD LOGUE and STEPHEN WEAR

THE HEALTH CARE INSTITUTION / PATIENT
RELATIONSHIP

Over the last quarter century, the field of bioethics has proceeded with a primary focus on the sorts of moral problems and quandaries that surface within the individual provider-patient relationship. To be sure, a parallel focus on more systemic issues, such as justice and equity within health care, or the allocation of scarce medical resources, has also occurred. Equally, focus has been given to more extraordinary, "sexy" sorts of issues, e.g., stem cell research and surrogate motherhood. But if one performed a tally of the number of pages produced within bioethics proper over this period, the total count for issues such as informed consent, truth-telling, confidentiality, beneficence and paternalism, and death and dying, cast within the specific frame of the provider-patient relationship, would surely dwarf the number of pages dedicated to the more systemic or extraordinary issues.

We do not mean to quarrel with this historical pattern; it was probably the only way that the substantial progress recently achieved in bioethics was possible. Truly, within this last quarter century, a profound sea-change has transpired in health care as it moved from the "silent world of physician and patient," (Katz, 1984) with its robust paternalism, to what has been called the "new ethos of patient autonomy" (McCullough and Wear, 1985) wherein informed consent is a routine activity, and truthfulness and candor is expected from providers. Equally, the routine overtreatment of moribund patients has been at least partially replaced by reliance on advance directives, routine withholding and withdrawal of life-sustaining treatment, and comprehensive end of life or hospice care. And all of this has been codified, as one sees in patients' bill of rights statements hung on the walls throughout health care institutions, reference to such institutions' various "ethics" policies, and as healthcare staff are reminded whenever preparation for the visit of the Joint Commission for the Accreditation of Healthcare Organizations (JCAHO) occurs (Wear, 2002).

We do intend to suggest, however, that this predominant focus on the ethics of the provider-patient relationship has become increasingly anachronistic in important regards. This is so simply because the primary relationship of patients within health care has essentially shifted from the relation they have with an individual provider to one in which their primary relationship is to a healthcare organization. The type of institution varies, of course, from HMOs to PPOs, to medical groups, to merged

Ana Smith Iltis (ed.), Institutional Integrity in Health Care, 99-110.
© *2003 Kluwer Academic Publishers. Printed in the Netherlands.*

hospital systems. But if we are looking for a current analogy to the constant presence of the old family doc, and the nature of his or her relationship with individual patients, we will less and less likely to find it in any one individual, but in some institution that regularly provides patients with their healthcare. As patients move from ambulatory/outpatient care, to acute or intensive care units, to nursing homes or hospice, the faces constantly change; only the organization remains constant.

This simple, overwhelming fact, we submit, calls for a basic re-consideration of many of the traditional ethical issues. For is not such continuity of care crucial to effective practices regarding informed consent, and death and dying? It is generally agreed that informed consent is a process that must occur over time, as is the development of a knowledgeable patient. But the development of such a knowledgeable patient surely becomes the responsibility of many different caregivers over time, if it is to be coherent and effective. Equally, many of the problems within the area of death and dying appear to have less to do with the generation of living wills, or health care proxies, by patients, and much more to do with whether the healthcare institution manages to identify such documents upfront and somehow insures that they are present at patients' bedsides, or in emergency rooms, when needed (Wear, 1998, see pp. 29-48). More basically, the bedrock of continuity and coherence of care for any given patient is only provided by an institution, through the hopefully organized and efficient interactions of its staff, and the quality and currentness of its documentation. The omnipresent family doc is just not around anymore to pull all this together.

1. THE CONCEPT OF INSTITUTIONAL INTEGRITY

We can think of no more appropriate notion to capture this new, primary reality of healthcare institutions than that of institutional integrity. Quick reference to any dictionary, in fact, shows the notion of integrity to be a very rich one, incorporating many of the traditional provider-patient relationship virtues. Its basic meanings include honesty, fairness, justice, fidelity, trustworthiness, truthfullness, candor, scrupulousness, veracity, virtuous, moral rectitude, excellence, fulfillment of one's duties, and dependability. Equally, toward the point that institutions can, and should, be moral agents in their own right, meanings such as totality, collectiveness, indivisibility, integrity, and unity may be found.

The above noted, however, one can quickly perceive that the notion of institutional integrity is just a new way of capturing an old series of problems. Institutions are obviously the sum of their parts and such parts can surely interact well or poorly. Simply consider record-keeping. The memory banks of the old family doc must now be replaced by what is captured in the patient's chart, and as the patient moves from provider to provider, and site to site, this institutional memory of the patient's history can be more or less adequate and available. Similarly, aside from the different parts of the institutional "whole" that staff represent, there are clearly numerous "stakeholders" in any such institution, such as: those who foot the bill at whatever level, providers, employees, patients, communities, and regulators, e.g. JCAHO, FDA, OSHA or state Departments of Health, and all of these entities may well have different views on and approaches to any particular issue or enterprise that the institution may confront. More generally, the unity, nay integrity, of any such institution can easily amount to no more than a Tower of Babel where its parts are pulling every which way, and only superficially seem to be proceeding with a common set of purposes and language. The potential for the numerous right hands not knowing what their counterparts are doing is vast.

A more optimistic approach here, however, is to take seriously the notion that what we all took to be at the core of the struggle to provide good, humane and ethical health care, i.e., the provider-patient relationship, has been eclipsed by a new, more primary relationship, that of the health care institution and its patient. Having made this jump, we can then proceed to address, again, the staple topics of bioethics over the past quarter century, as in: how to provide the best care while being responsible stewards of scarce medical resources; or how to assist patients to be knowledgeable about and participate in medical decisions where time, interest, capacity, and need for this vary markedly; or how to protect confidentiality where good care, as well as the protection of others, require broad availability of patient records.

There is surely an irony to this shift as reference to the older bioethics literature reveals that part of the impetus for the "new ethos of patient autonomy," was the post-WWII distrust of institutions. It was not clear to many that such institutions were trustworthy, it seemed obvious that they perceived people on the group not the individual level, and in no sense could one reasonably think of such institutions as having a conscience, exercising compassion and empathy, etc. Institutional integrity was thus more likely to be just an oxymoron, not a desideratum, and the antidote to this pathology was taken to lie in the informed/empowered patient sharing decision making with the humanistic provider.

But our point is that this is now the only game in town, and merits being taken as seriously as the provider-patient relationship once was. Let us first review a couple of basic, traditional ethical issues, viz., informed consent, and death and dying, to see how these might be recast or augmented within this new reality.

2. THE ETHICAL DIMENSIONS OF THE RELATIONSHIP BETWEEN PATIENTS AND HEALTH CARE INSTITUTIONS

That patients must be provided informed consent, be dealt with honestly and with candor, and their care and treatment be beneficial and humane, is expected. But how might health care institutions effectively step into the vacuum created by the absence of the old family doc? Let us first separately review the core issue areas of informed consent, and death and dying, and see what the basic problems and opportunities may be.

2.1 Informed Consent: Event & Process

The medical paternalism of the past certainly had its articulate advocates, an advocacy based on arguments as to what was best for patients (beneficence) (Wear, 1998, see pp. 49-63). The arguments in favor of the "new ethos of patient autonomy" seem to have won out, however, at least for now in the developed West, a result we applaud. But given the new reality of care by institutions, not specific individuals, the danger of falling back into a mindless sort of paternalism, a new "silent world of patients," surely threatens, and threatens in a way that neither good arguments, nor intentions, may be able to oppose. A patient can quite easily become, again, the "liver in room 412B4" in any health care institution, both because the knowledge of providers "rotating through" rotations, shifts or sites, is no more specific than that, and because the patient is given neither time nor assistance to present and see himself or herself as any more than that. What has been gained regarding patient autonomy and respect over the last few decades can easily be lost, even with no one intending it.

Half of the battle here must clearly lie in re-dedicating ourselves to the enterprise of developing knowledgeable patients who are ready, willing and able to share in medical decision making, become knowledgeable "consumers," and be more compliant and self-monitoring. As providers routinely change, as sites of care change, especially to a much greater reliance on outpatient/ambulatory care, the need for a knowledgeable patient becomes all the more crucial, no longer just some nice luxury that we may or may not have time to try to produce, and patients may be more or less comfortable with and adept at this. In effect, half of the response to the vacuum created by the absence of the old family doc needs to be supplied by patients themselves, however much the other half, the institution, tries to fill the vacuum from its own side.

The core of such an effort lies in the process of informed consent and allied educational efforts, along with the enterprise of empowering and encouraging patients to be more knowledgeable and involved, much less passive. We will presume to offer a few basic caveats in this regard:

2.1.1 Developing the Knowledgeable, Autonomous Patient
Patients vary markedly in their desire to be informed and participate in medical decision making. Often it is probably the case that their knowledge base is pretty scanty, if not inaccurate, even when they receive an otherwise decent informed consent, as all studies of the effectiveness of informed consent document (Wear, 1998, see pp. 49-63). For some time, we have all tended to live with this sort of result, e.g. the not very well-informed patient, and have tended to respect patient "preferences" and tendencies, even to the point of allowing them to waive informed consent, formally as the law allows. More subtly, but commonly, we have also tended to accept the patient who is just not very attentive and is clearly waiting to be reassured and fixed, not empowered and turned into a knowledgeable decision maker.

Now, as has been argued (Wear, 1998, see pp. 100-124), many medical decisions, arguably the vast majority of them, can be legitimately addressed by provider recommendations that can be legitimately seen as in most patients' interests, and commended on the basis of the quite similar values and goals that most patients have, e.g., sustaining life, thwarting or diminishing disability and pain, etc. So the inattentive patient, in the usual case, has no *immediate* need to be prodded into functioning as a knowledgeable decision maker. Our initial point here, however, is that what is not immediately problematic in the inattentive patient still misses an opportunity to educate and empower that patient. And given the above, that patient is then less able to assume their half of the enterprise of filling the vacuum presented to us. Our generic suggestion, then, is that we need to become much less accepting of the quite common, inattentive patient. Such "sensitivity" and "respect" does him or her no real favor. And to the extent chronic or potentially terminal illness threatens, the loss of such an opportunity is all the more unfortunate.

2.1.2 The Core of Informed Consent

This point leads to the allied issue of what an informed consent should actually be or should aim at. Reference to the literature shows much argument in favor of informed consent as an ongoing process, depreciating the significance of the informed consent "event," usually accompanied by some informed consent form (Wear, 1998, pp. 85-99). In part this should be seen as a false dichotomy as however much education needs to be a process, in a crucial sense a decision must occur at a given point in time and has its own reality and needs.

Having said this, however, it appears that we have bought into a sort of process of informed consent that is long on identifying the trees, but quite short regarding the forest. That is: consents tend to focus on individual procedures or therapies, e.g., a central line, or anti-depressants, rather than gaining permission to proceed with an overall plan of care. In the intensive care setting, for example, the need for placing a central line may be merely a routine step in an overall, anticipatable process of responding to serious illness, e.g., an exacerbation of COPD. But what the patient really needs to understand is what problem(s) he or she has, e.g., COPD, what its causes are, e.g., smoking, what can or can not be done about it, and what real choices are present, now or in some likely future. Thus when a "real" decision presents itself, e.g., the need for intubation, it should be seen as at least as important to see such choices within the overall context of the patient's medical history and the specific plan of care that the choice is arising within. Then patients may have a chance of responding to the real issues in a way that actually brings into play their "personal values, beliefs and life-experiences".

Such a "core disclosure" (Wear, 1998, see pp. 116-121), given the emphasis it merits as the "forest" the patient is in, is what we take to be the heart of legitimate consent and education, not the numerous "trees" that the informed consents focus on. Repeated focus on this core reality, with updates, encouragement of questions, *and* especially the potentially time consuming tactic of asking the patient to report what he or she understands his or her problem to be and the proposed response, with suitable feedback and correction, is thus all the more requisite, and ongoing education that much more crucial.

2.1.3 Enhancing the Institution's Knowledge of the Patient

The role of staff in encouraging, prodding and even insisting on such a development of the knowledgeable patient should be apparent. Absent this, half of the opportunity to respond to the vacuum created by the evaporation of the old family doc is lost. The other half lies in the knowledge readily available to the staff that is caring for the patient at any given time and here the key seems to lay in the creation of a much more organized, specific and available institutional memory of the patient. Progress notes, discharge summaries, and summaries generated at the point of major decisions or turnings in a patient's care need to become much more centered on and specific to what the patient was experiencing: what they were told and what they understood, whether they were ambivalent or reticent in some regard, etc. Often times much of the "physiological history" of the patient, which overflows in the usual chart, is mainly ancient history, and its absence not much of a barrier to current diagnosis and treatment. But the patient's experience and activity as a person and a patient, how they saw and reacted to prior illness and resultant health care, may by crucial in managing the patient later, particularly when major decisions obtain, non-compliance, ambivalence or reticence about care surface, or one is simply wondering "where the patient is at" toward connecting with them in some realistic, effective fashion.

The heightened availability and power of the electronic record is clearly of enormous potential advantage here. However well done previous paper charting was, much of it is effectively lost as paper charts are regularly stripped and sent somewhere "downstairs", and staff often does not quite find the time to locate them later. But key events in a patient's health care history can be separated out and made immediately available at the computer terminal, including past discharge summaries, and descriptions of the actual character and content of provider-patient interactions at major points of decision and response, all including, as suggested above, a much more patient centered focus to include what was said, responded, problematic, etc., and who said or asked what of whom. The potential here appears to be enormous, not just for responding to the vacuum created by the absence of the old family docs, but even for surpassing what they likely had in their memory banks, if truth be told.

2.2 Death and Dying

All of this becomes especially needful when we move to patients where catastrophic, chronic and terminal illness occur, and aggressiveness of treatment issues arise. As already noted, the real task here may lie much less in patients generating advanced directives and much more in institutions somehow managing to make such documents available at the bedside in a timely fashion, accompanied by a real, specific sense of the past history, experiences and interactions that produced them.

Equally, staff must be ready, willing and able to utilize and honor such documents and insights. Research consistently shows, however, that health care institutions are still consistently failing to do this routinely and well (SUPPORT Principal Investigators, 1998). The documents are generally not available and, even when they are, they are often not honored. Similarly, patients, their surrogates and staff often have not adequately anticipated and prepared for end of life scenarios that were as predictable in their occurrence, if not timing, as the next rising of the sun.

Our overwhelming experience, from having developed and served on numerous ethics committees, is that full institutional support and encouragement is absolutely essential if such problems are to be rectified. And all too often ethics committees are tolerated only to the extent that they satisfy JCAHO requirements and do not attempt to modify and enhance provider behavior. The institutional culture may well instead opt for placing risk management and the sensitivities of the lawyers before what might amount to a fair and balanced approach to ethics committee activity. Thus barriers, in effect, get erected to what should be facilitated, and staff quickly gets the sense that this is not a major institutional priority among all the other things that are. And conscientious staff end up struggling against the current.

But how specifically might institutions support, encourage and enhance good, ethical practice in this area? We believe it is important, first, to realize that much of what passes for solutions these days are rather inadequate. Specifically, the use of advanced directives to resolve problems at the end of life has been over-touted in its potential and, in reality, such documents often cause more problems than they solve.

2.2.1 The Deficiencies of Advanced Directives

All too often, advanced directives, either in the form of living will documents or prior verbal statements, are either (a) much too vague to offer any real assistance, or (b) too specific in their declarations and end up inappropriately foreclosing on real options that the patient should have, but probably did not, consider. The most extreme examples of such vague directives are when the patient indicates something about not wanting anything "extraordinary" if they become a "burden." But many seemingly well-crafted living wills provide little more assistance in that either they only describe extraordinary scenarios where aggressive care would be futile anyway, e.g., persistent vegetative state, or use terminology, like terminal illness, that is actually poorly defined (is one terminally ill if some very invasive, lower probability intervention has some significant chance of cure?), and really is not decisive for the complex decisions and scenarios that actually present themselves.

The contrary deficiency occurs in part in response to the preceding one. That is: if one really wants to avoid needless suffering at the end of life, certain potentially beneficial activities must be rejected as well to accomplish this. What we have in mind here is when patients reject certain treatment(s) across the board, e.g., ventilators, because they fear anticipatable bad deaths given their diagnosis; e.g., people with chronic obstructive pulmonary disease (COPD) who fear ending up stuck on a ventilator, but end up foreclosing the real, potential benefits of such treatments in the process, e.g., the possibility that the bronchopneumonia on top of their lung disease is an acute, eminently treatable infiltrate that merely requires a short term, trial use of a ventilator to allow antibiotics to cure it.

Part of the etiology here certainly arises out of patients who unwisely make such advanced directives, written or verbal, without consulting health care providers who might have lead them to more nuanced sense of the possibilities that merit addressing. This can happen because it does not occur to patients to seek providers' counsel in such cases and they end up generating such statements to family members, or to their lawyers, as when they do some estate planning and sign off on the lawyer's preferred living will form in the process.

Even with trained clinical assistance, however, the enterprise of anticipating and speaking to future scenarios is difficult at best. However much the activities proposed in the last section succeed in generating knowledgeable, autonomous patients, the past may offer no more than questions to raise, not answers to act on. Further, patients may well reject extraordinary care or conditions when still living relatively ordinary lives, but when it's a choice between that ventilator, or being in the ground, minds tend to change. As one pertinent Spanish aphorism has it: "as one moves from the stands into the arena, the aspect of the bull changes".

Such initial thoughts lead us to commend a certain summary view of advanced directives, which we can then attempt to place within the health care institution context that we are seeking to enhance. That is: living wills, touted a decade ago by some as the panacea for death and dying problems, merit at best a mixed review. Often either too vague or specific, they ignore both the fact that minds often change when people are faced with the fullness of decision, and the problem that what is being anticipated is as hazy now as it may well be crystal clear in the future. Our own recommendation here is for people to opt instead for designating a health care agent to speak for them when they are unable to do so themselves, talk to that agent before hand to the extent that they actually have specific views or sentiments regarding aggressive treatment, and keep written prior directions to a minimum. As a colleague recently put it: "talk more, write less."

2.2.2 Institutional Support and Enhancement for the Management of Extraordinary Care

We use the generic term "extraordinary" here to signal that the focus needs to go way beyond cases of clear terminal illness or profound catastrophes, e.g. persistent vegetative state. It should also include patients who are faced with debilitating chronic conditions, e.g., Multiple Sclerosis or Amyotropic Lateral Sclerosis, as well as simply those patients who simply do not desire aggressive care in any situation, as in otherwise health, elderly patients who make clear they do not want to be admitted to intensive care if simpler, less invasive modalities prove inadequate. Given the quite poor results of codes in the elderly, arguably any elderly patient should at least consider being a DNR, whatever else they might accept for acute, potentially reversibly disease. We are thus arguably talking about a majority of any hospitals' population, i.e., those who, if properly informed and asked, might well indicate that they do not want everything that contemporary health care has to offer, now or later.

Now beyond whatever assistance the above commended program to enhance patient autonomy and knowledge might provide, it should be crystal clear that the health care institution can make or break any such process, either by enhancing and supporting the activities of its staff in such difficult situations, or by thwarting them, passively or otherwise. The institution can itself end up being the greatest barrier to good, ethical care in this area, and in numerous ways: (1) by accepting the all-or-nothing mentality wherein patients must choose between being a full code/full treat, or being a DNR with comfort measures only, as if there is not a wide spectrum of possibilities for most patients. Many patients then end up, as noted, getting more than they would have wished, or foreclosing on possibilities that might have provided benefit. A similar result often occurs, we believe, when getting enrolled in a hospice program generally requires that one become a DNR (as if it were clear that one might not suffer an eminently reversible arrest) as well as reject most modalities that might extend life, or even provide a long shot chance of curing what may well be a terminal illness. All this may well be prudent, but there surely is nothing fixed about it and one might hope for a more nuanced and flexible approach, rather than an all-or-nothing one. (2) Being overly fastidious about the withholding-withdrawing distinction can skew all this in a major way, as when patients refuse extraordinary life support that might have had short term benefit because they perceive, often quite accurately, that once started, such treatment is very difficult to stop. True progress will have been made in this area, we believe, when people, clinicians especially, come to conclude that there is no moral difference between not starting something, and stopping it if it proves ineffective, and withdrawal of life support becomes routine and unremarkable.

How might health care institutions do better? Generically, institutions need to create a culture where end of life and other situations are approached with the same reflectiveness, comfort and shared decision-making as any other type of care. All too often, staff manage to "get the DNR" from the terminally ill patient and spend little further time reflecting on the numerous further management issues that will arise in such a patient. The complex, reflective approach to the treatable patient somehow evaporates when we flip into less than aggressive care management; but this is just ostrich in the sand stuff, protecting no one, and harming many.

More specifically, numerous strategies and tactics commend themselves in this area. To the extent that the previous recommendations regarding producing knowledgeable, autonomous patients are followed, a foundation will already be laid. Further, early contact with and discussions with patients' surrogates, including health care agents, *before* the patient becomes incapacitated, could develop relationships, enhance their knowledge base, and provide prior patient directions and sentiments for later use that are much more specific and closer to "real time." Finally, to the extent extraordinary, non-beneficial scenarios are eminently predictable, e.g. respiratory distress in an end stage lung cancer patient, specific, formal directives might be solicited from the patient.

Again, the electronic chart has enormous potential if used and developed properly. The possibilities here include: (1) In one institution we are affiliated with, discharge summaries, DNR and limitation of treatment orders, and ethics consults, all have a separate and prominent place in the patient's electronic record, existing either on, or a mouse click or two from, the cover page that is the first screen that arises when one calls up the patient's electronic record. (2) One could certainly add to such available, primary data by including specific next of kin forms, scanned-in advanced directives, or summaries of major patient and/or family meetings to discuss issues and provide education; (3) a comprehensive, electronic "DNR/Limitation of Treatment" form could be developed that (a) provides a drop down menu calling for the basic elements of any good progress note, including why the DNR made sense medically, the reasons the patient and/or surrogate consented to it, the findings regarding the capacity of the patient, and so forth; (b) a requirement that any DNR order must include a statement of whether the patient will receive full, aggressive treatment short of arrest, or that he or she will also have non-arrest treatment limited in specific ways; thus a DNR ceases to be vague poetry, instead a specific clinical order, and non-arrest management is specified upfront, not guessed about at crisis time; (c) a pithy list of non-arrest treatment modalities that may be ruled out upfront could be developed, from intubation, to surgery and dialysis, to artificial feeding and hydration, rather than leave staff guessing what the DNR means at two in the morning when the patient decompensates; and (d) provision for the patient to indicate whether he or she would want such limitations of treatment to also apply if and when he or she becomes an outpatient, or during a subsequent admission. All such information could be immediately available to all staff, including those responding to an emergency in a patient they have never met.

The electronic chart can also vastly improve institutional knowledge of the patient in other ways. At point of contact, an electronic form asking for the patient's next of kin, whether he or she has made out an advanced directive, and where the document is, could be developed. It could also include the option of asking patients whether they would like to make an advanced directive (especially a health care agent, given the above argument), and if answered affirmatively, an electronic alert could be sent to the receiving healthcare team for response. Further, being electronic, such information could be readily updated by any staff members who get new or corrected information, which often occurs post-admission on the floor. Anyone who has spent hours trying to locate patient surrogates at a critical time, often because the paper charts' information is dated or inaccurate, should welcome such activities. Equally, those who have proceeded without the assistance of advanced directives that the chart indicates exist, but somehow does not contain, will see the potentially enormous possibilities here.

3. THE CONSCIENCE OF THE HEALTH CARE INSTITUTION

Earlier in this article, regarding the idea that the primary health care relationship is now between patient and health care institution, we observed the following:

> There is surely an irony to this as reference to the older bioethics literature reveals that part of the impetus for the "new ethos of patient autonomy", was the post-WWII distrust

of institutions. It was not clear to many that such institutions were trustworthy, seemed obvious that they perceived people on the group not the individual level, and in no sense could one reasonably think of such institutions as having a conscience, exercising compassion and empathy, etc. Institutional integrity was thus more likely to be just an oxymoron, not a desideratum, and the antidote to this pathology was taken to lie in the informed/empowered patient sharing decision making with the humanistic provider.

The opportunity for skepticism is vast here. Just like opponents of Medicare in the sixties predicted the inefficient, resource profligate system we now have, one might suggest that the rise to prominence of health care institutions is simply part of the same pathology, the same wrong turn, that health care has taken over the last half century. And thus the argument would run that we do not need to try to fix what does not work, but find another alternative, one that somehow more directly empowers and re-connects patients with individual providers.

Fair enough ... perhaps. One might choose to re-emphasize and insist that health care institutions re-double their efforts to insure that individual providers and patients continue to connect with each other over time. Further, even as patients' illness evolves so that sub-specialists get involved, or treatment beyond the expertise of the primary provider must be utilized, e.g., intensive care, continued engagement of the primary provider might be insisted upon. There was a time, of course, when the patient's "family doc" followed the patient into the hospitals and retained more-or-less final say on whatever was done, whatever the views of consulting specialists. Now you rarely see them. But we see nothing necessary in this. And a corrective arguably needs to be systematically applied.

Still, we believe that there is enough "water over the dam" here to continue to insist that the primacy of health care institutions will remain a reality, and thus the notion of the "integrity" of such institutions can not be allowed to be an oxymoron. The routine disengagement of the patient's primary provider actually has a long history. Those who early on joined HMOs may well have noticed that such institutions made no great attempt to ensure that the same provider be present for one checkup after another. The facile assumption seemed to be that such providers were interchangeable and that little or nothing was lost in the transitions. If the above argument is at all correct, this is just false and corrections need to be applied. But still, particularly with the movement to treat and follow patients much more as outpatients, the caveats regarding developing knowledgeable, autonomous, self-monitoring patients certainly applies, and there appears to be too many advantages to outpatient care to forsake this strategy. Equally, as much that happens to patients is irregular, and thus not schedulable, any attempt to go back to continuity of care by individual providers will be at best partially successful.

Our point at this juncture is certainly that, given current realities, the notion of institutional integrity, in its fullness, must be taken seriously; too much is at stake to do otherwise. Further, even if the modifications suggested in the previous paragraph occur, our sense is that their nature and effect will be at best partial and the problems identified above will still require response on a systems level. Thus, health care institutions must somehow become worthy of trust, exercise compassion and concern, and actually have a knowledge base that penetrates to the level of the individual patient. Equally, such institutions must be the primary vehicle and impetus for the development of the autonomous, knowledgeable patient commended above, as well as change their culture to more adequately countenance and respond to problems in the area of death and dying.

In short, institutions must develop something akin to a conscience, a process whereby "best practices" are identified, instituted and then monitored for compliance by staff. Similarly, the process of identifying medical errors, or inadequate providers, or system structures that provide barriers to or undermine continuity and coherence of care, must be all the more aggressive and wide-ranging. The institution must constantly be investigating whether it is actually doing what it says it will do, not sweeping its glitches under the rug, as in the past. Quality Assurance (QA) or Performance Management (PM) programs must thus be like a conscience in attempting to identify departures, individual or systemic, from good practice, not just the local "cops" who respond to high profile sentinel events.

In the end, our sense is that such an emphasis and focus on health care institutions as ethical agents in their own right, as "entities" for whom the notion of integrity in all its fullness is a primary touchstone, is as unavoidable as it is crucial. We have attempted to suggest some specific ways in which institutions might convert this idea from an oxymoron to a reality, specifically regarding the development of autonomous, knowledgeable patients, as well as in response to the numerous and profound problems in providing extraordinary care.

State University of New York at Buffalo
Buffalo, New York, USA

BIBLIOGRAPHY

Katz, J. (1984). *The Silent World of the Doctor and Patient.* New York: Free Press.
McCullough, L. & Wear, S. (1985). 'Respect for autonomy and medical paternalism reconsidered,' *Theoretical Medicine, 6,* 295-308.
SUPPORT Principal Investigators (1998). 'A controlled trial to improve care for seriously ill hospitalized patients: The study to understand prognoses and preferences for outcomes and risks of treatments (SUPPORT),' *Journal of the American Medical Association 274(20),* 1591-1598.
Wear, S. (2002). 'Teaching bioethics at (or near) the bedside,' *The Journal of Medicine and Philosophy, 27(4),* 433-445.
Wear, S. (1998). *Informed Consent: Patient Autonomy and Clinician Beneficnece in Clinical Medicine,* 2nd ed. Washington, D.C.: Georgetown University Press.

STANLEY JOEL REISER

CREATING AN INSTITUTIONAL ETHICAL IDENTITY

Institutions became the dominant landmarks on the map of health care as the twentieth century drew to an end. This rise to power occurred without the presence of a substantive ethical framework upon which to base corporate actions. Instead, as they had for centuries, institutions relied mainly on the ethics of their professional constituents to project the appearance and/or provide the substance of a guiding ethical presence.

How hospitals and other health care institutions grew under these circumstances, the consequences of this development, and remedies for it are explored in this essay.

1. THE RISE AND SHAPE OF PROFESSIONAL ETHICS: HIPPOCRATIC BEGINNINGS

The origins of professional ethics in Western medicine are found in the writings of Hippocrates and his disciples, which were composed over the 200 year span of the 5th to 3rd centuries BCE. Of the ethical works in this corpus, however, the Hippocratic Oath has been most influential (Hippocrates, 1923). Indeed as one surveys the development of health care ethics, perhaps its key feature and primary source of influence and change has been its codes, composed to encourage specific behaviors and to distill the essence of ethical thinking current at the time they appeared. These codes both display and clarify the relationship between professional and institutional ethics.

The Hippocratic Oath is central to the development of Western medical ethics, both for the particular ideals of individual behavior set forth in it, and for a mechanism it creates to assure their effective use. Let us consider first the ideals. They are enumerated in the Oath's second half, beginning with a standard that asserts the dual responsibilities of physicians to provide benefits and prevent harms. The existence of a clear benefit is the necessary justification for a therapeutic action. However actions directed at help must be modified by attention to the preventive ideal of eschewing avoidable risks. A powerful theme in Hippocratic medicine was a scientific recognition of the power of nature and the relative limits of therapeutic means, which meant not using them in situations where success was unlikely. A high moral requirement of practitioners also was respect for life, embodied in stern warnings against the practice of assisted suicide or abortion. Other ideals in the

Ana Smith Iltis (ed.), Institutional Integrity in Health Care, 111-119.

Oath include the need to demonstrate trustworthiness to patients and families in dealing with the physical intimacies and personal confidences required in therapeutic interventions, and an egalitarian view of humanity that committed physicians to treat all in need of their care with the same dedication.

These values were unlikely to have a long-term influence on the actions of the students of medicine by whom the Oath was taken were it not for a crucial innovation proposed in the Oath. The innovation is a pledge taken by the students who swore to the Oath to accept a life-long obligation to support and maintain a relationship with their teachers. Ethically this had the effect of binding students to follow the standards and obligations of the Oath, which their teachers stood for and lived by, and thus of creating growing numbers of physicians who practiced under similar ethical ideals. Institutionally this produced what is now called a profession—a group devoted to complex technical actions undertaken with reference to an agreed-upon set of values.

The Oath as ethical code, underpinned by the lifelong student-teacher relationship, became an organizing nexus of practitioners and professionalized them.

2. THE ETHICAL INNOVATIONS OF THOMAS PERCIVAL

By the start of the nineteenth century, the ethical ideals of medicine were propagated among practitioners through Greek documents like the Hippocratic Oath, medieval ones such as Frederick II's *Medieval Law for the Regulation of the Practice of Medicine* (circa 1240) (Frederick II, 1977) and essays and books written by learned doctors like Samuel Bard's *A Discourse upon the Duties of a Physician* (1769). However they sometimes proved too general to address particular medical situations, a circumstance that led the British physician Thomas Percival to write a comprehensive treatise first published in 1803, titled *Medical Ethics*.

Percival was a physician at the Manchester Hospital who was asked by hospital authorities to help resolve a controversial dispute among colleagues. The experience led him and others at the hospital to conclude that an extensive and detailed work on ethical conduct was needed to help avoid medical controversies and to better resolve them when they occurred. His book went far toward this goal.

He wrote of the need for hospital physicians and surgeons to recognize the importance of their work and the need for moral rules to govern it. Relationships with hospital patients were to be founded on "skill, attention, and fidelity," with as much attention given to patient feelings and emotions as to symptoms. In the wards physicians were urged to take patient histories using a voice level that could not be heard by other patients. Physicians were to be attentive to religious needs of patients and to remind those who were seriously ill about the significance of a will.

Much attention is given to inter-professional relationships, warranted by the circumstances behind the treatise. The comments are specific and explicit, for example: "No professional charge should be made by a Physician or Surgeon, either publicly or privately, without previously laying the complaint before the gentlemen of the Faculty." Detailed recommendations about hospital procedure are given such as: seeking new remedies and surgical procedures when older ones fail; conducting no new trials without consulting other practitioners (a harbinger of modern Institutional Review Boards); and adopting a hospital register to record the number of admitted, cured, and deceased patients, their ages, occupations, diagnoses, and so forth.

However the work followed the basic pattern established in the Hippocratic Oath of making the physician the standard-bearer of ethics. The teachers in Hippocratic days, and hospital authorities in Percival's time could serve as the inspiration, source, or guarantor who helped assure the ethical principles were followed. Still it was not hospital authorities but each physician upon whose shoulders lay the full responsibility for carrying the principles out. The mantle of medical ethics was worn by the individual doctor.

3. THE AMERICAN MEDICAL ASSOCIATION'S ETHICAL CODES

This viewpoint was continued in a series of documents directly connected to the lives of contemporary physicians—the ethical codes of the American Medical Association (AMA). They began in 1847 with the publication of the first code, an influential document whose language mirrored the traditions of the Hippocratics and Percival. The writing of ethical standards to guide the actions of American physicians was one of the two main reasons for forming the Association, the second being the reform of medical education.

The first AMA code is divided into three sections. One specifies the duties of patients and physicians to each other, the second examines the relationship of doctors to one another and to the profession, and the last section explores the mutual responsibilities of the "profession" and the "public." While the American Medical Association is the source of these activities the institution *per se* assumes no explicit ethical duties—all actions flow through its physician members. This is exemplified in the last section. Although the first of its two parts is titled "Duties of the Profession to the Public," it speaks only of the obligations of the physician—to secure the community welfare, give aid to society in medico-legal matters, provide assistance *pro bono* to indigent patients, and discourage the sale and use of secret or quack remedies. What is more, the heading of the section's second part drops any reference to profession and is titled "Obligations of the Public to Physicians." It speaks of society's need to give respect and encouragement to physicians.

This personal rather than institutional emphasis continued as the AMA code was revised during its first hundred years. But an institutional presence emerges in two of the 10 articles in the code's penultimate 1957 version. The first article states that "the principal objective of the medical profession is to render service to humanity with full respect for the dignity of man" and the fourth one declares that "the medical profession should safeguard the public and itself against physicians deficient in moral character or professional competence" (AMA, 1957, p.1).

The difficulties confronting organizations in taking an active role in defining ethical responsibilities for themselves is revealed in the American Hospital

Association's (AHA) Statement on a Patient's Bill of Rights, composed in 1973. The sixteen years that had passed since the 1957 AMA code appeared had produced the beginning of a modern renaissance for medical ethics. By the early 1970s, the ethics movement was extending its reach into major population constituencies and health institutions in the United States, exemplified in this AHA document produced for patients by hospitals.

The document reflects the growing significance of the autonomy principle as the cornerstone of the new ethics movement, the rising status of the patient who was the chief beneficiary of this ideal, and the deepening recognition by the hospital community of the new authority of patients. However in the document's preamble, the AHA affirms the centrality of the doctor-patient relationship in medical care before it justifies its legitimacy to compose a list of patient rights: "The traditional physician-patient relationship takes on a new dimension when care is rendered within an organizational structure. Legal precedent has established that the institution also has a responsibility to the patient" (AHA, 1973, p. 4).

4. IN SEARCH OF THE HOSPITAL'S ETHICAL IDENTITY

This mild AHA challenge to the authority of physicians in structuring therapy was the leading edge of a storm that would soon threaten to swamp their hegemony over patient care. In fact the rise of hospitals had been meteoric in the 20th century. Scientific advances such as antisepsis that vastly elevated the safety of care in hospitals and the willingness of patients to be treated in them, the need for a site within which the increasing array of technologies such as radiological and laboratory equipment could be accessed readily by physicians and patients, the growing cadre of specialist physicians who needed to be in close proximity to patients and each other for effective care-giving, and the pride taken by communities able to attract a hospital were among the factors that caused a spurt in their numbers. From about 400 in 1875, the number of hospitals increased to some 4,000 in 1909 and to almost 7,000 in 1928 (Rorem, 1930, p. 9). By 1973, the year in which the AHA's patient's bill of rights was written, there were about 7,500 hospitals in the United States (Health Resources Statistics, 1976, pp. 339 and 342).

Until the mid-1960s, physicians exercised the principal decision-making authority in hospitals, dominating their two other branches of governance, the administrators and the board of trustees. The hegemony of physicians began to weaken, however, with the passage in 1965-66 of two unprecedented and major government health insurance programs, Medicare and Medicaid. By giving health access to large numbers of senior and indigent citizens, these programs not only brought to hospitals many new patients but also government involvement and oversight. The vastly expanded and more complicated relationship between the hospital and the government and its fiscal significance to hospitals gradually but inexorably increased the responsibility and power of administrators, whose knowledge of economics, accounting, and management was not held by physicians (Stockle and Reiser, 1992).

By the end of the 1970s, the expanding power of administrators was boosted by a fiscal crisis. During this decade health care costs, of which about 65% were related to hospital care, were rising at a pace about double that of the national rate. This led the main payors for health care, business and government, to encourage reforms that held the promise of providing more cost-effective therapy. In turn, this accelerated the movement toward hospital consolidation, which was directed at achieving greater profitability through the cost efficiencies that increased size and buying power made possible. Hospital corporations flourished throughout the 1980s, some having hundreds of hospitals under their umbrella.

In the 1990s, after the failure of the health care reform legislation of the Clinton administration, the concept of managed care and the organizations that embodied it took hold of health care. When he campaigned for the presidency and after taking office in 1992, President Clinton had labeled containment of health care costs and inflation as his most important economic goal. He declared that they affected not only the health of Americans but the health of the American economy. Thus when his plan was rejected by Congress, the country embraced the rescue offered by the private sector—managed care.

Essentially, managed care is a system of limiting costs by gaining the administrative authority over essential health care procedures, personnel and organizations. This involves routing patients enrolled in its health plans to institutions under contractual agreement to charge specified low prices for therapy. To reduce their prices and survive, hospitals and other care-giving organizations connected to managed care had to manage their human and material resources strictly. Wherever possible, therapists who made lower salaries were substituted as care-givers for those whose wages were higher (such as nurse-practitioners for physicians, nursing aides for nurses). Protocols were devised for treating different diseases, which sought to restrict where possible the use of high-cost drugs and devices and limit the number of days spent in the hospital. For example, many managed care organizations reduced the allowable stay for childbirth to one day. Thus by determining the place, length, type and time of treatment managed care succeeded during the 1990s in holding down health costs.

As the concept of managed care spread throughout health care it was embraced by hospitals, whether attached to large health corporations or independent. Now physicians were confronted by corporate structures exerting authority not only over institutional policies but the substance of their own practices and decisions. They had become just another medical factor, albeit an important one, to direct and oversee. This corporate hegemony caused the American Medical Association to endorse the formation of physician unions in 1999 in an effort to recover lost authority (Mangan, 1999).

5. THE USE OF ETHICS IN HOSPITAL SETTINGS

It is ironic that despite the growth of institutional power over health decisions and personnel, hospitals still continue to derive the ethics of their institution from the ethics of their clinical staff.

Yet each of the two largest associations of health organizations has an ethical code. The American Hospital Association's "Ethical Conduct of Health Care Institutions" was published in 1987. Its three parts deal with community, patient care, and organizational responsibilities. A laudable concern for the community is expressed in this statement but largely in terms of joining other health agencies and providers in the community to address its health status, needs, and knowledge. This focus is more an extension of clinical concerns about delivering services than about institutional policy. The section devoted explicitly to patient care promises to foster confidentiality, informed consent, and procedures to resolve conflicts with patients and assure quality of care. The part concerning organizational responsibilities makes cogent comments on having equitable employee policies and the need not to compromise staff efforts to provide appropriate therapies. But it was the violation of this last goal that mainly fueled provider and patient discontent with health institutions in the 1990s operating under managed care principles: they may have read the message in this document but did not heed it.

Largely in response to complaints from consumers and their political representatives about harmful constraints placed on treatment by managed care health plans, in 1996 the American Association of Health Plans published a "Code of Conduct." It urged health plans to provide enrolled patients with information about plan benefits, exclusions, and physician reimbursement; to discuss options for patient appeal of a plan's coverage decisions; to assure confidentiality of health information, choice of primary care physicians, access to specialty care, assistance when changing plans, and loosened constraints on paying for emergency care; and it embraced the institutional use of practice guidelines, quality assessment, and utilization management. While the code sought to increase patient information about health plans and choice within them, it still reaffirmed their commitment to managed care principles of control of personnel and material resources.

The code did not alter consumer or government reservations about managed care and did little to change its practices in the field. Thus opposition to managed care grew stronger in the years immediately following the code's publication. The focus of the opposition was a national effort to pass a patient's bill of rights, which would have given patients legal protections against inappropriate denial of needed therapy. It was nearly made into law in 2001 and continues to have wide public support.

From the content of these two codes, it is evident that health care institutions retain the perspective of their clinical component and depend on its ethical aura when they address contentious ethical issues and write of ethical obligations. To fathom why hospitals and other care-giving health institutions have not embraced the goal of achieving an ethical stance *per se*, whose content is focused on different issues from that of its constituent professional staff, one can point to the daunting power of a some 2,500 year old ethical tradition in medicine that seems able to provide a large enough umbrella to cover all ethical needs, an institutional emphasis that is focused more on the business aspect than the caring side of health provision, inadequate ethical training in the administrative education of corporate executives, and the failure of administrators and professional hospital personnel to recognize the separate ethical identities and needs of institution and staff.

An interesting example of these behaviors is found in the hospital ethics committee, the principal site of ethics discussion in hospitals. It originated in the early 1970s as the prognosis committee convened at the Massachusetts General Hospital to examine appropriate care for very sick patients, and the committee of hospital physicians required by the New Jersey Supreme Court to be part of the clinical decision process in the case of Karen Ann Quinlan, a young woman in a coma whose life was sustained by a respirator. Hospital ethics committees grew significantly in the 1980s to assist clinicians, patients, and families with ethical problems generated by clinical decisions. While they have been helpful with a number of institutional activities such as creating ethics education programs and crafting policies and protocols for hospitals in areas such as end of life care, and have administrative as well as clinical representatives on them, most of their time was and is spent on ethical problems involving the clinical staff. In their some three decades of existence there has been little evolution by them toward engaging the ethical problems of the hospital as an institution.

The need for hospitals to cultivate a separate ethical identity flows from their place in society as independent entities. The hospital is given a charter. It establishes a board of trustees and hires administrators to govern it. It is a legally independent agent that can be held liable for medical actions undertaken within it. It constructs mission and vision statements to state its goals. This entity needs to define and embrace ethical standards by which to measure its actions. It requires mechanisms to discern, discuss, and resolve ethical issues that institutional actions generate. It needs a culture of work structured around ethical values. It needs an ethical identity.

6. ESTABLISHING AN ETHICAL IDENTITY

An identity is who someone or what something is. For institutions devoted to the sacred calling of treating the sick and safeguarding the well, the ethical facet of their identity is crucial. To create this ethical identity I propose seven guiding ideals whose conjoined purpose is to infuse the institution with an ethical presence.

1. Validate staff worthiness. Fairness should influence the forging and implementation of institutional rules and procedures. Gratitude should convey the value attached to staff dedication and service.
2. Advance the welfare of neighbors. Each institution resides in a neighborhood and is a member of a community. This creates a responsibility to assure that their basic interests and well-being (including

but also going beyond health needs) are advanced and not damaged by institutional policies.

3. Respect the public. Constancy and honesty to the public about the institution's performance and goals are essential to gain and deserve its trust. The public should have a true picture of the institution's effectiveness in treating diseases and conditions, its standing when compared to like health institutions, the qualifications of personnel staffing its facilities, and the health outcomes it produces.

4. Serve society. Because health institutions have the implements and personnel to prevent and relieve human suffering and respond to the threat of death, they have a compelling obligation to strive to bring their skills to those who need them. Managerial efficiency joined to philanthropic responsibility are essential to creating the capacity and will in the organization to perform this service mission.

5. Accept responsibility to future generations. Institutions differ from individuals in their unlimited capability for longevity. A health institution which is successful at self-improvement has more to give each succeeding generation. This potential gift requires a stewardship of resources that ethically balances present commitments and future duties.

6. Create conditions that safeguard and benefit patients and students. Health institutions need capabilities to assure the exercise of best practice in carrying out therapeutic and educational responsibilities.

7. Promote a caring culture. Caring behavior should be uniformly exhibited in all sectors of the institution. A health institution that is uncaring to its staff cannot expect them to be fully caring to students and patients. The standing and treatment of residents, whose identity ambiguously straddles that of staff member and student, merit particular ethical attention.

Guiding ethical ideals need the presence of the administrative instruments to assure their salience in policy formulation and institutional life. One is institutional ethics rounds, which can focus on emerging issues to prevent problems, ongoing issues to solve problems, and past events to anticipate them. Ethical standards become empty ideals if they are not continually invoked in institutional life (Reiser 1991 & 1994).

Another way to encourage the use of institutional ethical standards is to establish a subcommittee of the hospital ethics committee devoted to institutional issues needing a special hearing and independent review. The subcommittee should work under the general rules of its parent committee in determining who should be given access, how findings and recommendations are used, where they are kept and so forth. As appropriate, the general rules should be modified to handle the requirements of exploring institutional matters. Having both clinical and institutional ethics questions dealt with by a common hospital entity encourages cross-fertilization of ideas and procedures between the two areas.

Ethical thinking should be a basic component of institutional health cultures such as hospitals. Now hospitals are bifurcated ethically into a clinical part that is educationally, historically, and socially committed to use ethics in its discourse and decision-making; and an administrative part that lacks these traditions and often finds it inexpedient and difficult to make a commitment to ethically-based policies. The ethical separation of the hospital's two basic communities is institutionally unproductive and perilous, since unethical behavior in one part invariably damages trust in the other. If these communities each base their actions on ethical grounds, they will have found the best common ground for a relationship that serves society and brings honor to the institution.

University of Texas Health Sciences Center at Houston
Houston, Texas, USA

BIBLIOGRAPHY

American Association of Health Plans (1996). *Code of Conduct*. Washington, DC: American Association of Health Plans.

American Hospital Association (1987). *Ethical Conduct of Health Care Institutions*. Chicago: American Medical Association.

American Medical Association (1846-1847). 'Code of medical ethics.' In: *Proceedings of the National Medical Convention 1846-47* (pp. 83-106). Chicago: American Medical Association..

American Medical Association (1957). *Principles of Medical Ethics*. Chicago: American Medical Association.

American Hospital Association (1973). 'Statement on a patient's bill of rights,' *Hospitals, 47*, 4.

Bard, S. (1769). *A Discourse upon the Duties of a Physician*. New York: A & J Robertson.

Department of Health, Education, and Welfare (1976). *Health Resources Statistics Health Manpower and Health Facilities, 1975* Washington DC: US.

Frederick II (1977). 'Medieval law for the regulation of the practice of medicine.' In: Reiser, Dyck, A.J. & Curran, W.J. (eds.), *Ethics in Medicine: Historical Perspectives and Contemporary Concerns* (pp. 10-12). Cambridge, MA: MIT Press.

Hippocrates (1923). 'The oath.' In: Jones, W.H.S. (trans.) *Hippocrates* (volume 1, pp. 299-301). Cambridge: Harvard University Press.

Mangan, K.S. (1999 August 6). 'Academic medicine becomes a target for labor organizing: physicians and residents seek union help on both economic and professional issues.' *The Chronicle of Higher Education*, A14.

Percival, T. (1849). *Medical Ethics, 3rd ed.* Oxford: John Henry Parker.

Reiser, S.J. (1994). 'The ethical life of health care organizations.' *The Hastings Center Report, 24(6)*, 28-35.

Reiser, S.J. (1991). 'Administrative case rounds: institutional policies and leaders cast in a different light.' *The Journal of the American Medical Association, 266*, 2127-2128.

Rorem, C.R. (1930). *Capital Investment in Hospitals*. Washington DC: Committee on the Costs of Medical Care.

Stoeckle, J.D. & Reiser, S.J. (1992). 'The corporate organization of hospital work: balancing professional and administrative responsibilities,' *Annals of Internal Medicine, 116*, 407-413.

CHRISTOPHER TOLLEFSEN

INSTITUTIONAL INTEGRITY

1. INTRODUCTION

To live well is largely to pursue various human goods in a fully reasonable way.[1] But a fully reasonable way is a way which is harmonious across several axes. A flourishing and reasonable agent is in internal harmony, is in harmony with the world, including cooperative harmony with the social world around her, and is in harmony with God. Culpable failures to pursue and maintain these forms of harmony are existentially crippling, and reasonable agents thus take these forms of harmony as themselves goods to be pursued even in the pursuit of other substantive goods such as health, knowledge or play.

What is true of individuals is also in various ways true of groups, institutions, states, and other forms of social structure. Social realities may be more or less internally harmonized, more or less at peace with their neighbors or competing societies, and more or less in harmony with God. Company workers, for example, may either comply willingly or unwillingly, or refuse to comply at all, with management policies. A nation may be at continual loggerheads with its neighbors, manifesting a state only nominally different from continued war, or it may provide aid of various sorts, forging bonds of friendship. And groups, companies, states, and all manner of societies may exist primarily, or in large part precisely to carry out purposes at odds with divine reality, or they may quite explicitly exist in order to serve God in all things.

Of course, internal, social, and religious harmony must be sought in pursuit of genuine goods. And debased forms of harmony should not be pursued at the expense of genuine harmony, nor otherwise legitimate forms of harmony be pursued when there is some real moral obstacle it. Peace with one's neighbors may not be purchased by systematically looking the other way, or, worse, aiding and abetting in their criminal enterprises. But still, the various forms of harmony are real goods not just for individuals, and they constitute in large part a society's flourishing. Just as for an individual, a society that maintains these forms of harmony in pursuit of genuine goods may be said to manifest integrity.

Integrity, then, is a crucial notion for our normative evaluation of social groups of various sorts. Failures of integrity are, by their nature, failures of the well being of a society; there can be no genuine trade off between the integrity of an institution and its flourishing.

Ana Smith Iltis (ed.), Institutional Integrity in Health Care, 121-137.
© *2003 Kluwer Academic Publishers. Printed in the Netherlands.*

In this paper, I pursue the concept of integrity with a view to understanding its role in the life of a social organization. This, in turn, is done with an eye towards understanding what the demands of institutional integrity are for two particular types of organizations: Catholic universities and Catholic health care facilities. As we shall see, not all institutions demand harmony to the same extent. In this paper, by distinguishing between strong and weak institutions, I explain this difference in the scope of an institution's demands on its members. But Catholic institutions of every sort, and especially Catholic universities and health care institutions, demand integrity and harmony across a very large scope. They are, in my terminology, strong institutions. Their integrity is thus, I shall argue, very easily damaged, and damage is likely to be very grave. But this is to anticipate.

The remainder of the paper has four sections. First, I sketch the various forms of harmony important to individual integrity and well-being. Although integrity properly so called essentially involves internal harmony, external disharmony can be a threat to integrity, or a symptom of lack of integrity. And the pursuit of goods which involve external harmony can be vocationally important, so as to shape an individual's integrity. In the second section, I discuss the integrity of institutions, following the model of the first section. In the third section I discuss religious and Christian societies generally, before moving on in the fourth section to discuss Catholic universities and healthcare institutions.

2. THE NATURE OF INTEGRITY

Human beings are complex; if we were angels, or pure substances, we would not be faced with as many possible ways of coming apart as we in fact are. We are rational, volitional, and passionate; each of these aspects of the human being may come into conflict with the others. Thus, Smith judges that it is best to give blood, but, perhaps because of an inordinate desire not to experience pain, does not choose to give; there is a lack of harmony between judgment, on the one hand, and choice, action, and emotion on the other. But even if Smith does indeed choose, and act, it may be that she still feels a strong inclination away from the object of her choice. We should, *pace* Kant, recognize this conflict as a form of disharmony between reason and choice on the one hand, and emotion on the other.

Moreover, Smith's various emotions may be in internal conflict; and she may not have considered well various aspects of her life, resulting in conflicts between judgments and commitments. By contrast, if Smith has made wise structuring choices, is of good character, and has educated her passions adequately, then we will think of her as a person of integrity: in internal harmony across the various aspects of her person.

There is a tendency to see integrity as possible for agents of bad character – surely, some think, a Nazi, or a Don Giovanni may be of high integrity, with judgment, will, and emotion all operating as one.[2] In one sense this is true – very corrupt characters are unified in their corruption. But in another sense it seems false; the reasonable life is a life spent in structured pursuit of some genuine human goods and openness to all. But immorality involves in all cases some form of hostility or closedness with respect to basic goods. Those who attack human life, or friendship, or art, or the good of marriage, have for that reason, avenues towards the fullness of human flourishing closed to them; they reject, in various ways, the good, limiting their opportunity for human flourishing.

In consequence, any integrity that is purchased through immorality is bought at the price of a radical limiting of the scope of opportunity for human well-being. But this seems to obtain unity in the wrong way for integrity. Consider the integrity of a work of art. An excellent painting or story has its integrity through a unity in multiplicity – multiple parts are structured into a pleasing whole.[3] A single speck of paint, or a single word on a page, has a kind of unity, but of a stunted sort; we would not say of such a work that it has integrity.

One aspect of human well being is found in society – in friendships, marriages, and sociable and cooperative relations with others. This diverse good is itself a form of harmony, so again, human well-being is damaged if this form of harmony is rejected, denied, or damaged. But, further, social goods, such as marriage, can provide a vocational focus for agents, around which their life centers. This vocational focus is central to the shape that integrity will have for such agents. This requires a brief explanation, as it will be crucial later.

A good life is, we have seen, fully reasonable, not just at a time, but through time. This requires that a good life take on a certain structured shape. The contours of this shape are provided by those commitments an agent makes to pursue and foster some goods and projects, rather than others – to get married, rather than remain single, to marry this person, rather than that, to accept children lovingly, and so on. Or, to become a philosopher rather than a doctor, to pursue this branch of philosophy rather than that, and so on. Such commitments are more or less architectonic, playing a structuring role in the agent's life, raising new demands, but also limiting other considerations that would have been important given other commitments.

It follows that vocational commitments are central to an agent's integrity, for reason, in most cases, judges what is appropriate for an agent in light of these vocational commitments. But many of these commitments – marriage is a central, but not the only, example – are social in nature. So the forms of harmony one pursues socially are often centrally defining of what integrity would be for an agent.

However, in addition to this positive role, the intersection of the individual and society can create enormous challenges to personal integrity.

These challenges exist because of the general lack of moral rectitude found in human beings and societies, conjoined with the fact that *societas* is a basic human good. For the immoral and irresponsible purposes and instrumentalities shared by many create pressure on individuals who wish to maintain harmony with the social world around them. A husband urges his wife to lie on taxes about part time work she has performed; a business encourages its employees to engage in minor acts of fraud for the company well-being; colleagues in a university strongly discourage faculty from speaking or acting in ways consistent with their faith. In all these cases, it is tempting to pursue a fraudulent form of sociability – the mere appearance, rather than the reality – by acquiescing to the demands made on one by society. In this way, society threatens an agent's integrity. I will return to this problem in the next section.

The final form of harmony available to human beings, and constitutive of their well-being, is harmony with the transcendent source of all meaning, God. This form of harmony is interesting and crucial because of its tendency to play a vocationally defining role in an agent's life.

This role is normatively demanded of all Christians, for example, whose lives are supposed to be shaped throughout all their concerns and pursuits in relation to this fundamental option – a commitment to put God first in all they do. Violations of this commitment result in radical disruptions of integrity. So the martyr, in refusing to deny God, not only continues to give to God what is due, but also, in that very act, maintains her integrity in relation to her most central and important commitment.

As we saw in the discussion of morality and integrity, we see here also room for a distinction between the full or deep integrity of a flourishing Christian, and the thin integrity of a nominal Christian.[4] Take, for example, a rather minimal commitment to the Christian faith, which made few demands – Mass on Easter and Christmas. It would not be difficult to maintain integrity here, but it would be a mere form, normatively and eudaimonistically insufficient in light of the full demands and promises of Christianity.

A final consideration with regard to religious commitment, harmony, and integrity. We saw earlier that social demands for harmony with others provided not only opportunity but also challenges to internal harmony and integrity. The same will be true of the demands of the social in relation to the religious aspects of one's life. This will be important in considering the ways in which strong religious institutions can fail.

3. INSTITUTIONAL INTEGRITY

There remains much more to be said of the integrity of individuals. However, it is necessary to move on. In this section, I show that societies, broadly construed to include institutions and organizations, may be assessed in much the same way as individuals. In response to a possible objection, however, I then distinguish between strong and weak institutions.

What is crucial is that societies be complex in ways analogous to individuals. The claim that such analogies do indeed hold, and that they are relevant to assessments of institutional and individual integrity, is old – it is found in Plato, to whom my account is much indebted.

The first form of integrity and its lack that we saw in the individual was the possible gap between what the agent considered to be demanded, and what the agent actually did. Temptation to do other than what, say, morality demands, can come from emotions not integrated with reason, and choice can acquiesce in emotions' unreasonable demands. Much the same can be true of institutions.

For this to be the case, societies must have structures for deliberation and decision, and indeed this is typically the case, features that are manifest in explicit organizational features – a board of directors, for example. These deliberative features exist precisely so that a society may consider and pursue its good, and so that the deliberations and decisions will have authority for all those in the society. But as with individuals, the good for a society is not always easy or pleasant. Apathy, greed, or squeamishness may lead those with deliberative and decision authority to chose an option other than that demanded by reason. Similarly, the recognized repugnance of the most reasonable choice to those subject to authority in the group might lead those in authority to choose some more palatable but less reasonable option and to act accordingly. In this way the institution can lack harmony of this first sort.

Likewise, there is always a difference between those with authority for deliberation and decision, and those charged with carrying out the decision. This is true even where decision requires unanimous agreement by all – those who have made the decision still must now carry it out. But a society may reveal one degree of disintegrity when those under authority refuse to comply; and another when, while complying, the workers are distressed by or unhappy with the decisions of those in authority. On the other hand, those under authority may be fully complying, not just with the letter of the decision, but with its spirit – understanding, agreeing with, and being pleased by whatever decision was taken.

From the analysis of section two, there may be internal conflict paralleling the conflict of emotions in a single agent. Not all in a society agree with a decision perhaps. Or, perhaps the institution's conception of its own ends, and the commitments around which it is structured, may not be well ordered, or may be positively disordered. An incoherent constitution, setting out ends which are not easily reconciled to one another, threatens institutional integrity. Concerns for this sort of integrity are found in discussion of the conflict of interest generated by certain for-profit approaches in health care.[5]

This last point brings out the way in which a society's constitution — written or unwritten, explicit or implicit — functions in a way analogous to an individual's vocational commitments. The constitution of a society specifies what will count as that society's common good, the good pursued by all members of the society in common, albeit in different ways. The constitution is identity-determining. The constitution thus shapes the contours of an institution's integrity – commitment to some goods demands certain choices that otherwise would have been optional, or even wrong, had other commitments been made. It is worth pointing out, perhaps, that since societies last longer than the lives of their individual members, unlike individuals, who last as long as they last, the constitution of a society plays a crucial role in allowing it to maintain integrity over change of membership and across generations. This is analogous to the way vocational choices structure an agent's life through time.

Consider, next, the relationship that societies may stand in to other societies and to the divine. Just as individuals may take certain social goods to be vocational hubs, so can institutions – health care institutions, to take an obvious example, have the health care of non-members as their structuring good, a complex commitment to the goods of health and solidarity. Here again, commitment to a social good plays the vocationally structuring role in a society's "life". It is thus precisely through the failure to serve the health of others that health care institutions suffer most damage to their integrity.

Similarly, some institutions, having the service of the divine as the, or a primary focus, flourish or suffer in their integrity precisely by their relationship to the divine. A church concerned primarily with money or reputation, that fails to provide adequate pastoral guidance to its members, suffers from a failure of integrity.

Nor, as suggested earlier, should it be surprising to find that the promise of harmony or threat of disharmony between an institution and the surrounding social world can provide serious threats to an institution's or society's internal harmony. An institution, like an agent, may go along just to get along; a religious organization can, for fear of seeming too confrontational, fail to speak out on crucial issues. Finally, in consequence of this preceding form of temptation, or for some other reason, a religious institution's commitment can become "thin", in which case the demand of integrity may be met, but the form of integrity become unhappily weakened.

Now, a crucial dissimilarity between individuals and institutions might jump out at us. While there is some opposition to the view that individual lives should be strongly unified, most find this an acceptable demand of morality. Even those aspects of a life that might seem out of step with a strongly unified character usually are, in fact, budgeted for in an agent's self-conception and commitments. Spontaneity, for instance, might be something an agent values, and thus not out of character.

By contrast, in many cases, it is undesirable and often wrong for an institution to demand so high a degree of internal harmony. Consider all the ways in which what the members of a business do, think, and feel are not just unregulated by the business, but at times at odds with positions taken by the business, or by other members of the business. It seems to be a moral requirement that such wide latitude be permitted: it would be offensively intrusive for a business to check, for example, political party membership of its employees. But this high degree of heterogeneity does not seem to threaten a society or institution's integrity as such.

The reasons for this latitude are various. Most obviously, there are limits, given basic human rights, to the extent that any institution can control various private aspects of its members. But more interesting in our context is a different kind of limit, viz., those limits drawn on the ends pursued by an institution. An institution's ends are limited, and they are intended to constrain member behavior only within this limited scope. This seems to be essentially a consequence of two related factors. First, societies do not properly have lives of their own; thus their flourishing is not valuable in itself, but for the sake of their members.[6] The members of any society, whatever its purpose, recognize that at least in some respects they are limited in their ability to pursue certain human goods, and they recognize further that cooperative action will be empowering in their, and others', pursuit. So for various limited purposes, societies exist so that by flourishing, those societies may benefit their members in regards to their limited purposes. But, second, and at the same time, very much of what constitutes an individual's flourishing is dependent upon what the individual does in spheres which must be left untouched by the institution. An individual's well being, for example, often depends upon successful personal relationships; but these by their nature must be pursued by the agent herself, and not by an institution on behalf of the agent. So the society must leave the agent free in this area.

In consequence, the common good of a society is usually only weakly specified: the society's goods are instrumental, limited, and do not require the full participation of members. But some institutions have strongly specified goods – they make considerably more demands on their members, tolerating less heterogeneity of action, belief, and desire. "Strong" in this context means that the demands will be more extensive, or that they will entail more rigorous demands on the shape of members' character, or beliefs. Why? In what follows, I discuss four kinds of circumstances that lead societies, groups, institutions, and so on to specify more strongly the common good of their members. Strong specification leads, in ways we will see, to stronger demands and thus to richer forms of integrity.

First, some institutions pursue only instrumental goods, but others pursue basic goods. In general, societies that pursue basic goods will make stronger demands on their members. So, within the limits set by ordinary morality, a society formed for the pursuit of some instrumental good will not make many particular demands beyond that of efficiency in some limited area, whereas institutions formed to pursue a basic good will rightly expect, for example, some stronger form of commitment to the good in question.

A second, slightly different, consideration concerns how narrowly or broadly and how shallowly or deeply the pursued good is specified. This distinction cuts across the previous one. An institution or group can pursue a basic good narrowly or shallowly: play, for example, but only inasmuch as the production and racing of model cars constitutes a form of play. Or it can pursue a good broadly and with depth, as various health care organizations do. Groups that pursue instrumental goods can likewise pursue very limited objectives – the production of this particular drill bit – or much broader and deeper instrumentalities, as, for example, the complex of instrumentalities pursued by the state. Broad or deep pursuit of goods will result, other things being equal, in stronger demands.

Thirdly, there can be an intersection – sometimes a necessary intersection, between the goods pursued by a society and the vocational structures of its members. Some institutions are established precisely to enable individuals to fulfill their specific vocations, rather than to enable them to establish subsidiary aspects of their vocations. A consequence will be a stronger tie between the individual's flourishing and the flourishing of the society, a tie not present in many weak institutions. A member of an unsuccessful club, or business, for example, might himself flourish, but a monk in a dysfunctional monastery is one's whose flourishing is by that very fact under serious threat.

Fourth, it might be the case of some institutions etc. that one of their purposes is to provide witness to the goods pursued, rather than simply to pursue them. It is arguable, for example, that a university exists not merely so that its members can pursue truth, but also to stand as a witness to the value of the pursuit of truth. Institutions of this sort are strong. Similarly, the goods pursued by a group or institution might be reflexive. That is, they might be such that their successful pursuit requires that their members be willfully committed to the pursuit or those very goods as good, rather than merely material cooperators in the venture. This will be true of religious institutions in the sense in which I define them below.

I am quite sure that this is not a complete taxonomy, nor is it as fully analytically realized as it could be. But it does suggest that there is a considerable range of demands that groups, societies, institutions etc. will legitimately make on their members, and a considerable range of commitment, character, and belief that will be acceptable from within the institution.

It follows that the scope of opportunity for robust integrity, or failure of integrity of the group is considerably enlarged in strong institutions. Institutional integrity is a matter of there being harmony within an institution between that part of the society responsible for judgments and decisions, where these judgments and decisions are in turn structured by the overall purpose of the society and its common good, and those charged with carrying out decisions. But since, in weak societies and institutions, the common good is narrowly specified, the range of judgments and decisions over which there must be harmony is correspondingly limited. In strong societies, by contrast, the common good demands more; thus more is required for institutional integrity.

We have, I would suggest, a secular model of what strong societies are in many professions. Professions seem to come into being when, for the sake of more effective pursuit of important social goods, institutions are established to promote the skillful pursuit and distribution of those goods. Professional devotion to these goods is often described as "vocational". And by their behavior, professionals are expected to witness the value of what they do, even in their private life. Hence the shame of a prospective head of the justice department being discovered to have violated tax law.[7]

4. THE INTEGRITY OF RELIGIOUS INSTITUTIONS

Many religious societies, and especially many Christian societies, are strong on the account given above. And some specific types of Catholic societies – universities, health care institutions, and religious orders in particular – are very strong.

It is natural that religious institutions should tend towards strong specification of their common good and of the demands made upon their members. For religious institutions and societies in many cases meet every criterion listed above.

What I mean in this context by religious institutions, societies, and groups, is organizations whose primary purpose is defined in terms of its relationship to the divine. Because of the nature of the understanding of what that relationship calls for, the primary purpose of such organizations will most likely be defined in terms of service, for it is through service to the divine that we expect to be pleasing, and thus able to enter into friendship.

Service to the divine for the sake of harmony with the divine should not be understood only in terms of the form of prayerful communities entered into by some monks or nuns. For the architectonic reason for which agents pursue various other important and basic goods — health or truth, for example — can also be as a form of service to God. This is more clearly the case when, as in many Christian societies, God is served by serving those in need, precisely insofar as they have needs related to basic human goods. But in these cases, human beings and human goods are served not simply instrumentally, but both in themselves, and insofar as such service is pleasing to the divine.

This brings us to the first point mentioned above: religious societies serve basic, non-instrumental goods, most significantly the very good of religion. But even, as we just saw, when God is served by serving man, man is served vis-à-vis basic goods, such as life, health, education, and so on.

Secondly, many religious organizations pursue their specified goods in broad and deep ways. The depth of pursuit of the good is evident in many religious societies. Many form the institutional framework, for example, within which the bulk of an agent's life is to be spent in increasing focus upon the good and its pursuit; this contrasts with, for example, a club, which provides a context in which goods are less fully pursued. And breadth is evident in the fact that such societies do not propose specific goals or states of affairs to be achieved, but rather establish patterns of life by which their good or goods are to be participated in.[8]

Even more obvious is the way in which religious, and especially Christian and Catholic societies and institutions, exist for purposes that overlap significantly with individual vocations. As mentioned above, the common good of a monastery is not detachable from the particular vocation of each monk; a religious order devoted to serving the poor must have as its members persons whose vocation it is to serve God by serving the poor. Such institutions do not go wrong in mandating, for example, that a long period of testing and reflection be undergone before an agent is allowed to fully participate in the institution in order to ensure that this vocation is indeed present.

Finally, religious societies and institutions, especially if they are Christian, are explicitly charged with bearing witness to their purpose, the service of God. Christian institutions are described as being "evangelical". By this is meant not only that they carry the word of God to others, but also that they reveal the truth of the word in all that they do.

Three further points of clarification may be offered here. One is to reiterate what was implied earlier: an organization, institution, society, etc. is not religious simply because many, most, or all of its members are religious, or even because it calls itself "Catholic" or "Christian". What I here mean by religious societies is, again, societies formed precisely to serve God as their primary purpose.

Second, there is a clear distinction between weaker religious organizations — a club, a Bible study group, a rosary group, for example — and the sorts of strong religious societies that are the focus of this discussion. But what will characterize this difference is primarily the lack of depth and breadth pursued by clubs and small groups, and the lack of full intersection of vocational structure and group purpose. Indeed, in most such cases, the primary overlap between a religious society and the individuals' vocational structure will be found in the individual's presence in the Church as a whole. A Bible study group is formed in order to provide an instrumentality by which individuals' various vocations in that larger body, the Church, may be pursued.

This allows us to anticipate, and fend off, a possible objection to what I will shortly discuss, namely, the ways in which Catholic universities and health care institutions are strong. For in opposition to this claim, regarding, for example, Catholic universities, one might point to some Catholic club, and the way in which such a club is weak by my standards, making few, or non-rigorous demands on action, belief and character, and demanding little in the way of conformity. But, I shall argue, the nature of the goods pursued by universities and health care institutions, and the ways such goods must be pursued, are such that when pursued for the sake of service to God, they can only result in very strong institutions. The ways by which such institutions become weakened – as for example, when a university ceases to require doctrinal orthodoxy of its theologians – can occur only by the university sacrificing its original overarching apostolate.

Third, a consequence of all this, and related to both previous clarifications, is that the primary purpose of a strong religious institution or society can never, whether explicitly, or merely in practice, be reduced to any goal or state of affairs, nor its primary measure of success be efficiency in the pursuit of such goals or states of affairs. The most effective relief agency, if it is, or was intended to be a religious relief agency, may be a complete failure, for neither the primary good of a religious society, nor the basic goods that are subsidiary to that primary good, are states of affairs or goals.

Finally, the connection between this discussion and the overarching issue of integrity should be emerging. Strong religious institutions must do a number of things to maintain their identity, and to preserve it at all levels. Judgments must be made pursuant to the society's true good; decisions taken that reflect best judgment, rather than external or internal passions or desires; and those subject to authority must, in all the relevant ways, be one with those with the authority.

Thus, to take one example of a way in which a religious institution's integrity is threatened in virtue of the special demands its nature makes, consider a religious hospital that refuses to perform certain forbidden procedures, such as abortion. Does the hospital manifest integrity simply because no one in the hospital in fact performs such services, even if some, or even many of those subject to the hospital's authority believe abortion to be a legitimate and sometimes necessary procedure?

I find it difficult to believe that such a hospital manifests integrity. For if the hospital is indeed to provide witness to the evil of abortion and the value of human life, and to witness that in serving the least protected of human beings it is serving God, there must be willingness on the part of all who might come into contact with a patient desiring an abortion to speak and otherwise act in a way that reflects that willingness. Moreover, it should be expected that nurses, for example, when asked about the hospital's stance while off-duty, should be ready and willing to defend it, rather than unwilling or disparaging. The witness provided by the hospital does not extend merely to its walls, but to what its members do throughout the rest of their lives.

5. CATHOLIC HOSPITALS AND UNIVERSITIES

Why is it then, that Catholic universities and hospitals must either be strong religious institutions, or non-religious institutions? And what specific consequences may be drawn from this?

To a certain extent, it should be clear that universities and hospitals are by nature moderately strong institutions. This is reflected in the thought that medicine and the pursuit of truth and the education of the young are professions. Truth and health are basic goods, not, in any setting, easily reducible to narrow pursuit of limited states of affairs. The understanding of the nature of these goods, essential for their successful pursuit, is the task of a lifetime – they are thus vocational hubs for the lives of many individuals who pursue them. And because they are genuine goods for all, and of crucial significance to the flourishing of all, professionals in these fields are expected to bear witness to the importance of these goods. So, for instance, an academic's plagiarism is viewed not just as a narrow failing in regard to the good, but as having a strong social dimension. Such an academic fails to bear witness to the importance of truth and casts the profession into shame.

Still, in various ways, secular universities and hospitals are relatively weak. This is most clear in the secular university. Because of widespread diversity of belief, religious background and so on, academic freedom is taken to be something of an absolute value. But rather than strengthening the demands made on faculty, academic freedom tends to loosen them, allowing great divergence of view, even divergence as to the value of the overarching value of academic freedom.

There is perhaps a parallel here in medicine, which, in a secular context, tends towards a thin conception of its ends, focusing on technical interventions and limited therapies, as well as attention to patient autonomy, and abstracting from, or neglecting the good of the person as such. It must be admitted that there has been much criticism of modern secular medicine on precisely these grounds, that medicine must be humanized.[9] I am sure this is true, but at the same time, it is intelligible how, given the same sorts of diversity that affect universities' conception of their own good, medicine as well should increasingly take a narrower, more technical approach, abstracting from substantive conceptions of the human good.[10]

In addition, in both secular universities and hospitals, actions are likely to be performed and positions taken with which some employees, perhaps of a religious nature, for example, will be in disagreement. Within fairly wide limits, however, it is not thought that employees who continue to serve in such universities and hospitals in ways not directly linked to the questionable acts and positions must thereby sacrifice their integrity, nor that their disagreement threatens substantially the integrity of the institution. Clearly this does not mean that such employees should not attempt to reverse bad decisions and policies.

Finally, there are many who work for secular universities and hospitals who serve in only rather instrumental positions, who are not expected to have any necessary commitment to even the most general values of the institution. Cafeteria workers, cleaning and repair staff, secretarial staff — in many cases it will seem entirely acceptable that such employees should have no essential connection to the mission of the institution. Such employees are often considered interchangeable — efficiency being their primary virtue, others may take their place without loss, provided only that they can do the job.

Each of these situations seems considerably changed, however, in a religious university or hospital. The changes will be intelligible consequences of the much stronger specification of the common good that emerges when the primary purpose of an institution is specified as religious. This stronger specification occurs in at least the following ways.

First, the understanding of the basic goods served is deepened. In Catholic universities, no longer can there be the extreme divergence of beliefs characteristic of the secular university (extending, today, even to the belief that there is no such thing as "truth"). As John Paul II writes, "A Catholic University's privileged task is to 'unite existentially by intellectual effort two orders of reality that frequently tend to be placed in opposition ... the search for truth, and the certainty of already knowing the fount of truth'" (1990).

Likewise, in Catholic hospitals, the understanding of the good of health and the good of the patient is shaped and reinterpreted in light of Christian teaching on the dignity of the person and the relationship of all persons to God their Creator and Redeemer. This represents the deep grounding for the Christian call to "rehumanize" medicine, a call made, as mentioned, even in secular medicine, but more fully intelligible and gripping in the Christian context.

In both these contexts, moreover, understanding these goods in light of Church teaching sharpens awareness of the moral norms governing pursuit of these goods, as well as sharpening the understanding of the necessity of these norms. Catholic teaching against abortion, for example, will be seen not as simply a more or less onerous restriction, but precisely as the manifestation of the Church's great love and respect for all human life. Additionally, in the health care context, there is available an understanding and appreciation of suffering and its value that is simply unavailable from a secular standpoint, from which relief of suffering is often seen as the greatest good, to be pursued even through killing.

Second, the pursuit of truth and of health is in a sense transcended in Catholic universities and hospitals by their orientation towards God, and by their pursuit being now undertaken as a specific form of Christ's ministry. In this new context, the operative vocational hub, and the witness that is given, change; while the goods of truth and health, respectively, serve as vocational hubs, and are given witness, the primary vocation is service to Christ, and the primary witness is to the good news, good news that manifests itself in special ways in pursuit of these goods. Speaking to consecrated men and women, but, I believe, with wider ramifications for all those working in a Catholic health care ministry, John Paul II writes that it is "a part of their mission to evangelize the health care centers in which they work, striving to spread the light of Gospel values to the way of living, suffering and dying of the people of our day" (Pope John Paul II, 1996, no. 83).

Third, as mentioned frequently in Church teaching, those engaged in Catholic health care and education have a particular obligation, flowing from their sharing in Christ's ministry, to "the poorest and most abandoned." This obligation clearly has ramifications in health care – a preferential option for the poor manifests itself in indigent care, for example, although not, perhaps, as much as it should. And it should be present in the life of a Catholic college or university, in the provision of educational services to the poor, the elderly, the disabled, or in, for example, educational exchanges with institutions from less advantaged countries.

Now, the primary theses of this paper may be summarized as follows. First, given an understanding of the primary purpose of Catholic universities and hospitals that flows from the previous discussion, this understanding, internalized and accepted within such institutions, constitutes the core of the Catholic identity of a Catholic university or hospital. Second, this Catholic identity constitutes the basic self-understanding that will govern deliberation and decision-making among those charged with these tasks. Third, decisions made in light of this self understanding will be both broadly and deeply demanding on members of the institutions in question. And fourth, such broad and deep demands are unlikely if not impossible to be met in a way that demonstrates integrity in the institution, unless a significant number of members of the institution are practicing, devout Catholics. I consider the first three theses to have been generally illustrated and defended thus far. In concluding this paper, then, I make some claims as to the plausibility of the fourth.

Consider first the stronger conception of the good and the richer set of demands made in religious, and especially Catholic, universities and hospitals. These seem to have as an obvious consequence that the majority of those whose primary purpose is the service of these goods must be Catholic. For, as we have seen, the institution must not only serve the goods in richer ways, but for the sake of service to God, and in such a way as to bear witness to this service. This simply cannot be done in an institution primarily staffed by non-Catholics or non-Christians.

This is doubly so because the form of integrity demanded by this richer conception of the good extends, in a way that it does not in secular institutions, to a fuller form of harmony between the good of the institution, the decisions taken to serve that good, and the willingness of these employed to serve the good. While we do not think that the integrity of a secular university or hospital is seriously jeopardized by having numerous staff who disagree with some or other institutional policies, in a Catholic hospital, to have many staff who disagreed with the magisterium on matters of faith and morals would constitute a serious lack of institutional integrity.

The fruits of this failure of integrity can be seen in recent attempts to require Catholic hospitals to provide contraceptives in their insurance coverage. One argument made is that since the majority of employees in many Catholic hospitals are not Catholic, they should be provided with health services they deem desirable even if the Church disapproves. The argument is specious, it seems to me, but the difficulty seems to be one that Catholic hospitals have invited in their increasingly overriding concern to stay afloat whatever the cost.

So it seems plausible that those employees whose vocations overlap with the vocational identity of the institution should not simply overlap, but should cohere with that identity. That is, Catholic hospital and university workers should not simply be vocationally committed to the pursuit of truth, or the good of health, but to Christian service and witness to God through the pursuit of truth and the good of health. In Catholic institutions, this requires observant Catholic membership.

Earlier, I suggested that one difference between Catholic and secular institutions might be the extent of commitment necessary in those employees whose vocations did not overlap with that of the institution. That is, many employees in a secular hospital are employed only to perform instrumental tasks, necessary for, but not constitutive of the vocational life of the university or hospital. Only instrumentally do those who clean toilets or cook in the cafeteria serve the goods of truth or health.

There simply does not seem a rich opportunity for such workers to share in the commitments of faculty or doctors. For to the extent that for such employees their work is vocational, it is a different vocation than that of the institution at large, and the institutional purpose simply does not play into the actual vocational choices of the workers – they are not committed to the pursuit of truth or health; typically, they are cleaning toilets to make a living to support families. Apart from fulfilling their responsibilities, and with qualifications concerning good labor relations, they play a very minor role in the institutional identity and integrity of a secular school or health care facility.

At a Christian hospital or school, however, all employees can potentially share in the overarching vocation of the institution. For all employees can take service to God as their overarching vocational commitment, and do all that they do in service to the Divine. Moreover, all persons in a Catholic hospital or school can fulfill that commitment in a way proper to the institution itself. That is, a Catholic janitor at a Catholic university does not simply have a Catholic identity that overlaps with the membership of other Church members. Rather he can specifically contribute to the Church's mission in the university, by praying for faculty and students, by attending, in common with other community members, mass and prayers, and by bearing witness to the Catholic good served by the institution in representing the university to others.

This form of integrity is no doubt supererogatory; there can be many reasons, including significant reasons of charity, for hiring non-Catholic janitors. But it strikes me that the form of integrity available here is very desirable. If possible, why should it not be pursued?

On the other hand, the demands of institutional identity that the majority of those vocationally involved in the Catholic university or hospital be Catholic seems to be not simply desirable, but obligatory. There would be no Catholic identity without this kind of majority Catholic presence. To the extent that, today, most Catholic health care institutions and many Catholic universities are not primarily staffed by observant Catholics, such institutions seem to suffer from one of at least two forms of deficient integrity: either decisions are made for the true good of the institution, but are either not followed at all (e.g., when dissent is taught without disclaimer in a Catholic theology class) or are followed but only unwillingly and even with obvious signs of distress; or the decisions demanded by the common good of a primarily religious social reality are simply not made. There are too many stories of Catholic health care institutions that have entered into agreement with other institutions in order to survive which involve, e.g., the provision of forbidden services on "independent" floors of Catholic hospitals.

This treatment hardly exhausts all the issues involved in either the general topic of institutional integrity or the specific issues of Catholic universities and healthcare institutions. But I wish to close with three brief points.

First, I suggested earlier that the pursuit of harmonious relations with the outside world could operate as a severe temptation to violate institutional integrity. I believe that this has indeed been the case in both Catholic hospitals and universities, which have become increasingly conformed to the world. In some cases this is straightforwardly so that the institution will not suffer financially, in other cases, it results from the false promises of, secular status and prestige that Catholic universities seem particularly susceptible to desiring.

Second, both forms of institutions might be inclined to preserve their integrity in the face of external demands and temptations, not by choosing in obvious conflict with what their common good demands, but rather by paring down, or thinning, their identity, their conception of the goods they were pursuing. Health care facilities might think now of themselves as primarily in pursuit of the good of health, universities as primarily in pursuit of truth, but not in possession of it. And the understanding of these ends might further be whittled away in the way earlier described as characteristic of secular institutions.

But in this case, that is what these Catholic hospitals and universities have become. They are now more truly secular than Catholic, and it is a new kind of failure of integrity – the sort that arises from dishonesty – to continue calling themselves Catholic.

So, then, third, what should a Catholic institution do that sees its choices as follows: maintain integrity but fail to survive, or either violate the demands of institutional integrity, or thin those demands out so as to be followable? In addressing this sort of question, Germain Grisez remarks: "In bearing witness, individual Christians are expected to sacrifice even life itself when that is necessary. Should not [a Catholic order] be reading to bear witness by giving up its hospitals and finding other ways of carrying on its apostolate under today's changing conditions?" (1997, p. 400). I believe this is a suggestion that must be taken very seriously. If my reflections in this paper are on track, it is a suggestion to which considerations about Catholic institutional integrity seem inevitably to lead.[11]

University of South Carolina
Columbia, South Carolina, USA

NOTES

1. See Finnis (1980); the first two volumes of *The Way of Our Lord Jesus Christ:,Christian Moral Principles* (1983); *Living a Christian Life* (1993); and Finnis, Boyle and Grisez (1988).
2. Lynn McFall raises this as a problem for philosophical accounts of integrity (1987).
3. For an elaboration of this broadly Aristotelian view, see La Driere (1957 and 1959).
4. Or, as Ruiping Fan might say, of an anonymous pagan (Fan, 1999).
5. See Pellegrino (2000).
6. See the provocative remarks by Manuel Velasquez on the totalitarian consequences of thinking of corporations as being essentially superorganisms with persons as parts (1991).
7. I address the nature of professions at more length in Tollefsen (2002).
8. See, in this context, two remarks of John Paul II on consecrated life, in his *Vita Consecrata* (1996): "There is a youthfulness of spirit which lasts through time: it arises from the fact that at every stage of life a person seeks and finds a new task to fulfill, a particular way of being, of serving and of loving...consecrated persons must therefore be helped...to renew their original decision, and not confuse the completeness of their decision with the degree of good results" (no. 70). The Pope here is speaking specifically of middle aged consecrated persons, but in such a way as to show that throughout life their must be deepened awareness of perfection of one's apostolate, and that this should not be confused, narrowly, with the achievement of some state of affairs.
9. Edmund Pellegrino and Pope John Paul II have been especially noteworthy in this regard. See, e.g., Pellegrino (1977). See the see the various documents on health care of John Paul II at http://www.chausa.org/MISSSVCS/JPII.ASP
10. The most persuasive proponent of views of this sort is H. Tristram Engelhardt, Jr. See Engelhardt (1996).
11. The reflections in this essay took their root in considering Grisez's summary of John Paul II's teaching that Catholic hospitals "...must not only deliver quality health care but provide service to 'the poorest and most abandoned of the sick,' give religious instruction and encouragement along with health care, explicitly evangelize, strive to humanize medical practice, fully conform to the Church's moral teaching, and supply sound formation in that teaching" (Grisez, 1997, p. 393). I have tried here to supply some reasons for thinking these really are requirements for, e.g., Catholic hospitals. Grisez suggests, as I quote above, that in present circumstances, these requirements might demand "martyrdom" from many health care institutions. In an earlier essay, I suggested another possibility (see Tollefsen, 2000, p. 278.) I am now rather more inclined towards Grisez's view.

BIBLIOGRAPHY

Engelhardt, H. T., Jr. (1996). *The Foundations of Bioethics, second edition.* New York: Oxford University
     Press.
Fan, R. (1999). 'The memoirs of a pagan sojourning in the ruins of Christianity.' *Christian Bioethics, 5,*

232-237.

Finnis, J. (1980). *Natural Law and Natural Rights*. Oxford: Clarendon Press.

Finnis, J., Boyle, J. & Grisez, G (1988). *Nuclear Deterrence, Morality, and Realism*. Oxford: Clarendon Press.

Grisez, G. (1997). *Christian Moral Principles: Difficult Moral Problems*. Quincy, IL: Franciscan Herald.

Grisez, G. (1993). *Living a Christian Life*. Quincy, IL: Franciscan Herald.

Grisez, G. (1983). *The Way of Our Lord Jesus Christ: Christian Moral Principles*. Chicago: Franciscan Herald.

Pope John Paul II (1996). *Vita Consecrata*. Boston: Paulist Books and Media.

Pope John Paul II (1990). *Ex Corde Ecclesia*. Boston: Paulist Books and Media.

LaDriere, J.C. (1959). 'Literary form and form in other arts.' [On-line] Available: http://www.propylaean.org/litFormOther.html

LaDriere, J.C. (1957). 'Structure, sound, and meaning.' [On-line] Available: http://www.propylaean.org/struc_sound_mean.html

McFall, L. (1987). 'Integrity.' *Ethics*, *98*, 5-20.

Pellegrino, E.D. (2000). 'Economics and ethics: The right ordering of conflicting paradigms.' *Philosophical Inquiry*, *22*, 1-16.

Pellegrino, E.D. (1977). 'Humanistic base for professional ethics in medicine.' *New York State Journal of Medicine*, 1456-1462.

Tollefsen, C. (2002).'Managed care and the practice of the professions.' In: Jones, J. & Bondeson, W. (eds.). *The Ethics of Managed Care*. Dordrecht: Kluwer Academic Publishers.

Tollefsen, C. (2000). 'The Importance of Begging Earnestly,' *Christian Bioethics* 6(3), 267-280.

Velasquez, M. (1991). 'Why corporations are not morally responsible for anything they do.' In: May, L. & Hoffman, S. (eds.), *Collective Responsibility* (pp. 111-131) Savage, MD: Rowman and Littlefield.

DUANE M. COVRIG

INSTITUTIONAL INTEGRITY THROUGH PERIODS OF SIGNIFICANT CHANGE

Loma Linda University's 100 Year Struggle with Organizational Identity

1. INTRODUCTION

Integrity is a word denoting adherence to a certain value base or code of behavior. Identity denotes the distinguishing character or personality of an individual or group. Integrity and identity are connected because integrity is faithfulness to core identity through periods of change. This paper is a sociological case study of Loma Linda University's (LLU) identity and integrity through about 100 years of rapid change from 1905-1998. As a Seventh-day Adventist (SDA) healthcare and educational institution in Southern California, LLU made and still maintains simultaneous commitments to health care, religion and education. This chapter explores the development of those commitments and their influence on institutional integrity. This chapter grows out of a larger study of Loma Linda University's organizational development (Covrig, 1999).

In this chapter I first review my basic assumptions about institutional integrity and outline my methodology. Next, I chronicle key organizational developments in LLU's one-hundred year history. I then try to interpret these changes using sociological literature on organizations. Fourth, I draw some findings and discussion from LLU's example. Finally, I conclude with suggestions for the wider discussion and research on institutional integrity.

2. ASSUMPTIONS AND METHODS

It is my belief that administrators must continually work to maintain an organization's identity and preserve its integrity. Unfortunately, maintaining status quo is not the same as maintaining identity or integrity. A much more dynamic process is at work that requires choices that welcome change. Organizational inertia is insufficient to maintain integrity. Integrity is about actively maintaining a distinguishing character or personality, an identity, in the presence of competing possibilities for new directions for change. As such, identity and integrity are as

Ana Smith Iltis (ed.), Institutional Integrity in Health Care, 139-174.
© 2003 Kluwer Academic Publishers. Printed in the Netherlands.

much about maturation and transformation as they are about consistency and continuity. Integrity requires and even welcomes organizational changes that promise to better underscore the organization's core commitments and bring those values to bear more effectively on the world around them.

The challenge for organizational members and leaders is to figure out which changes violate identity and integrity and which changes help preserve and expand it. This is not an easy task for individuals within the organization and even more challenging for groups within the organization. Individually, each member faces the challenge of clearly isolating the core organizational values he thinks reflect the organization's core identity. Each needs to discriminate these values through a shroud of personal biases. Those biases continually work to morph the purpose of the organization into a form that will most likely match their personal bias and will most benefit their own skills or goals. Individuals subtly re-construct the meaning of the world around them to match their own internal view. Individuals do not believe what the see, they typically see what they believe (Weick and Sutcliffe, 2001). They will try to "wrap" organizations around their own interests and needs. Healthy reflection is the only way to prevent unhealthy distortion. It requires extensive emotional honesty and time and the progressive movement through stages of reflection (Kondrat, 1999). Such reflection is a rarity in most environments, including health care, where real internal and external organizational demands pressure individuals into blind and submissive obedience to policy or tradition. Even if individuals could escape their own personal biases and pay the high costs of reflection, they would still need to resolve the follow up on the painful discoveries of reflection.

Corporately, the challenges are even more complex on the road to institutional integrity. The hiring of each new member expands the possible interpretations of the organization's mission and identity. Vision and value pluralism make it difficult to maintain a unified vision or at least smooth operations between multiple understandings in an increasingly diversified organization. Administrations can avail themselves of several processes. They can work to reduce the uniqueness (identity) of the organization to make it more palatable to a widely accepted (generic) value frame. Or they can work by fiat and regulation to impose a view of the organization on members. This can be done by making employees sign statements of beliefs or commitments. The former process of compromise more often occurs in what institutional theorists call the work of mimicking (taken-for-granted) processes. The latter processes are regulatory institutionalism. Regulatory changes can be made by administrative fiat or across a sector of organizations through accreditation and regulations. I discuss these complex processes later in this chapter.

As difficult as it is to understand the complex individual and group processes of organizational integrity, it is even more challenging to sociologically research such processes. I faced four challenges in doing so. First, data needed to be collected from a long period of time to improve the odds that I would capture the essence of the organization and its choices in institutional integrity. Unfortunately, when longer periods are studied, a researcher must select data sources that are uniform across time. Organizations may experience many changes in their lifetime: changes in size, location, structure, names, mission statements, budget allocation, sources of support, types and quantities of clients and employees, forms of leadership, as well as the nature and extent of inter-organizational relationships. Tracking these changes was difficult through interviews. As such, documentary analysis was used. Formal documents never tell the whole story but typically are uniform. Thus, board minutes, policy statements, university newspapers, etc. were used in this study.

Such artifacts admittedly do not capture all aspects of organizational life but provide a uniform description of the organization. To off-set "paper" data, observations, and interviews can be used. Because I was a student for two years and a faculty member for three years at LLU, I was able to validate some of my documentary findings with participant observation. I also have been a lifelong SDA. I also formally and informally interviewed individuals. One individual had been at the institution since 1910, as a child of an administrator and later as a student, faculty member, and administrator. These experiences helped to clarify my document analysis.

Once data collection was handled, I faced the challenge of reducing the story to a readable format. Gathering a century of data into one chapter required great reduction. The challenge here was then to provide enough details to create an accurate portrait of the organization without overwhelming the reader with details that obscured the main themes. Unfortunately, authors and readers vary in their tolerance for details. Some want more data. Some want less. The history represented here is reduced from thousands of pages of artifacts and an extensive two hundred-page report on LLU (Covrig, 1999). I hope I have preserved enough of the story to be accurate without being tedious.

The third challenge once data collection and data reduction were handled was to use sociological theory to provide a theoretical backdrop to the "story" to allow the story to inform a wider field of organizational research. Such interpretative weaving is the great joy, pain and responsibility of scholarship. However, just as history becomes what historians say it is, so organizational identity and integrity often become what organizational researchers write it to be. To help me weave correctly, I had several researchers and LLU participants review this chapter for errors and faulty interpretations. However, what remains is my own biased scholarship, my own story of someone else's story.

Finally, the fourth challenge was analyzing institutional integrity by using both social science (descriptive) and philosophical (prescriptive) methods and statements without violating the scholarly protocols of either field. I am a novice at such work. However, I have tried to mimic two organizational researchers who have been masters at border crossing between the observed and the ethical (Selznick, 1992, 1996; Weick, 2001). My goal was not merely to be descriptive nor prescriptive. It was to be first descriptive and then form some philosophical statements about organizational integrity that would inform ethicists as well as sociologists.

In short, this chapter is my document-based, historically nuanced, socio-philosophical, interpretative analysis of the identity-creating choices that shaped LLU's institutional integrity.

The next section summarizes LLU's history through six arbitrarily assigned periods. Key events, choices and changes are described in each period. These events are linked to the organization's overall identity development. The next section of this paper uses sociological literature on organizational development, especially institutional theory, to nuance LLU's history. The fourth section offers findings and discussions about LLU. I conclude with suggestions on how this chapter might expand discussion about institutional integrity in health care.

3. LLU HISTORY

LLU developed from a small sanitarium-health resort in 1905 to a regional academic health care center by the early 1970s. LLU's history can be split into two phases. In the early phase, 1905 to 1950, LLU was a small college that primarily trained SDA health care professionals, some of whom went into international medical missions. In the late phase, from the 1950s to the present, organizational restructuring, a move to comprehensive university status, programmatic proliferation, a name change, increased diversity and popularity of health care occupations, rapid population increases in Southern California, and changes in leadership made LLU a major health science training center in the Inland Empire region of California. By the 1990s, LLU was attracting as many non-SDA students as SDA students. These two phases of LLU's development can be divided into six shorter periods (three in each phase) to show LLU's identity and integrity development.

3.1 Early Phase: Period One: The Founding Period, 1900-1915

The early 1900s to 1915 was the founding period in which a confluence of contingency factors and strong institutional values crafted LLU's founding identity. These forces involved *macro* processes such as national trends in health reform and Westward migration. They also included *meso* processes such as early Californian experimentation in health resort development and the emerging passion of SDAs for health missions and service. Finally, they included *micro* processes in LLU's local environments. Those included early SDA leadership decisions to support LLU, the early and powerful support of Ellen G. White (a religious leader of "prophet" status in the SDA community who retired to California), the availability of inexpensive resorts in California for SDA's to purchase for their emerging mission work, and the ability of LLU to attract inexpensive, talented, and committed workers as well as enough clientele to help cash flow. This convergence of multiple influences helped LLU avoid what Stinchcombe (1965) called the "liability of newness" (p. 148), which refers to the difficulty of a new organization surviving its formative years because of limited financial support and legitimacy. I will cover several of the influences that made a difference for survival and for identity formation during this period. I provide more detail for this period because the founding identity in this period is a reference point for later identity changes.

LLU was founded in 1905 as the College of Evangelists (1905-1909) and later became the College of Medical Evangelists (1909-1961). As these names suggest, LLU had a strong religious emphasis in its founding as it was established to promote SDA religious and medical ideologies and practices. Many of these ideologies and practices were hammered out only a few decades earlier in Battle Creek, Michigan. The SDA church officially started in 1863 in Battle Creek and the by the 1890s, even though it was still a small denomination, it had developed a three pronged ecclesiastical, medical, and educational organization. All three of these institutional aspects of the SDA church play into LLU's development as a church-owned, healthcare educational facility. The Battle Creek connection was important for LLU's early survival. A little diversion is necessary here.

The Battle Creek Sanitarium was probably the most famous item for SDAs in 1900. It prospered under the leadership of John Harvey Kellogg in the 1880s and 1890s. By the 1890s, Battle Creek Sanitarium had such a growing reputation that it soon was overshadowing the ecclesiastical and educational work of the SDA church. By the 1890s, the Battle Creek Sanitarium was sending out more medical missionaries than the ecclesiastical arm of the SDA church. These missionaries were starting small health centers around the world. However, by the 1890s, John Harvey Kellogg's successes and ideology created a schism between himself and the SDA church leadership. There was a growing distrust with Kellogg's "non-denominational" approach to medical work and training. This schism widened between Kellogg and SDA church leaders around 1900. Some leaders started to encourage Adventist's to move their operations out of Battle Creek. The small church college moved about 70 miles West toward South Bend, Indiana to the farming town of Berrien Springs, Michigan. The church's world headquarters moved to Washington, D.C. The training for medical doctors stayed in Battle Creek for a time, but as more sanitariums started up across the nation and world, nursing education was decentralized to those locations (Gerstner, 1996; Numbers, 1994; Schaefer, 1995; Schwarz, 1970, 1972, 1979, 1990a and 1990b).

LLU was founded in 1905 as a sanitarium, nursing school, and evangelistic training facility. At that time it was founded in a fairly unpopulated area near San Bernardino, about 60 miles east of Los Angeles. Within a few years interest emerged to make it a training facility for physicians. Distrust of Kellogg's work in Battle Creek had reached the point that many SDA church leaders were willing to support another medical training facility. As such, LLU started a training school for physicians in 1909 and eventually secured financial and governance support from the SDA world headquarters in Washington, D.C. by 1914, transferring ownership from local Adventists (Covrig, 1999).

Although directly funded and supported by the world SDA church, LLU, unlike Battle Creek Sanitarium, did not have SDA ecclesiastical leadership "down the street" to monitor day to day operations. The ecclesiastical leaders were across the continent. This gave LLU greater liberty to develop a religious entrepreneurialism in applying SDA beliefs to its own operations and challenges without close and daily criticism from SDA leaders. My reading is that this church-college relationship worked well for both institutions. It appeared to develop a mutual respect among many LLU faculty for their SDA heritage. Having financial support from the SDA church was crucial to LLU. Having a legitimate medical facility was important to SDA leaders. Even to this day, within SDA circles, I hear the phrase "West coast theology" to refer to the slightly different way LLU applied SDA theological ideas and religious practices. The culture of the West and the secular influences of medicine may have helped to explain LLU's "western" adaptation of SDA beliefs and practices. This is speculative, but it does indicate the structural unlinking of the center of SDA medical work from its ecclesiastical center.

Nevertheless, despite this entrepreneurialism, SDA religious ideas and practices had a profound and sustained influence on LLU's founding. The SDA doctrine of the human soul was one such influence. This doctrine rejected the theological view that the human soul was immortal. It argued against the prevelant dualism of the soul as separate from the body. This SDA doctrine fed into the SDA emphasis on health. It reinvigorated Christian thought about the the importance of caring for the body because it was the temple of God (1 Corinthians 6:19). SDAs believed, and still believe, that maintaining the health of the body is a vital spiritual duty. The medical and practical importance of this is clear. Care for the human body was a form of spiritual care. Restoring physical health was as much a part of the work of Christ's redemption as tending to the soul.This SDA belief was clearly emphasized in LLU's early public relations. I believe this doctrine gave LLU a strong bridge and entry into the hearts of SDAs, and helped it garner faculty, money, and legitimacy.

A second strong religious influence on LLU came from Ellen White. SDA's believed that White had the gift of prophecy, the ability to understand God's unfolding work for human salvation. Because she wrote voluminously, her ideas had a constant and steady impact on SDA thinking and action. This was true even after her death in 1915. When alive, White repeatedly said and wrote that LLU's founding was part of God's plan for SDAs. That simple support had a profound influence on LLU recruiting and fund-raising. Furthermore, White claimed that SDA views of health care and health care ministry came from heaven. This had an additional influence in motivating SDAs to live healthy lives and pursue careers in medical work. Although later medical historians demonstrated that many of White's ideas had a common origin in the American culture of her day (Numbers, 1973), the early view that SDA's had a special medical work to do gave LLU great legitimacy. White's direct support of LLU, including her dedication of the funds from her important book, *The Ministry of Healing*, gave LLU a founding identity that was rooted in religious experience (White, 1905).

Early supporters of LLU had unique religious and medical ideologies to promote through their work with LLU (Covrig, 2001b). While the spirit of medical sectarianism would later moderate within the LLU community, because of advances in medical knowledge, its still remains a strong source of legitimacy for LLU within its broader SDA community.

In summary, LLU crafted its founding identity from SDA religious doctrines and a growing religious passion to do medical work as spiritual mission. Rapid westward migration in America after the 1880s and the SDA conflict in Battle Creek in the 1890s set the backdrop for LLU's emergence. LLU's geographical start in rural Southern California would fit its early commitment to health resort identity and natural remedies. Even though it had traditional medical care (surgery and pharmacology), these identities would emerge more rapidly in the period to come. The region itself would grow rapidly in the 1900s. While this founding period set an early identity for LLU, environmental changes linked to rapid population increases, medical science discoveries and changes in post-secondary education, would alter that founding identity in profound ways.

3.2 Early Phase: Period Two: The Establishment Period, 1915-1930

Once founded, LLU faced a second identity-forming experience—becoming a legitimate medical training facility fully accreditated by the American Medical Association. During the next two periods, 1915-1930, and 1930 to the mid-1940s, LLU focused on meeting the demands of this new identity-forming constituent. In the establishment period (1915-1930) LLU utilized the people, resources, and legitimacy it had gained from its SDA community to fund facilities and program growth to AMA demands and serve a new area, Los Angeles. It was forced to expand into this area in order to expand its acute and critical care services. To earn greater medical legitimacy, LLU engaged in four activities. First, it continued the simple process of attracting patients, students, faculty, revenue, and legitimacy. Second, it expanded beyond its original ideological commitment from a rural "natural" health care center to include more traditional acute care in a city setting. The AMA required acute care for medical accreditation and LLU started a branch facility in Los Angeles to generate enough clientele to secure acute cases. The start of this clinic proved critical to LLU's development as a regional medical education powerhouse. Third, LLU mastered the art of using one organizational commitment (to the AMA) to frame its response to another organizational commitment (to the SDA). LLU was a religious institution, a point it utilized in clarifying itself to its medical community demands. LLU was a medical facility, a point it used in response to SDA leadership who were continually concerned about its actions and its uniqueness as an SDA institution. In subtle ways, having dual commitments gave LLU an ability to craft a unique identity between the demands of two competing interests. Finally, LLU secured the desired "A" rating from the American Medical Association and used that to attract more external support and wider respect among SDA and non-SDA medical workers.

The greatest identity-changing force on LLU during this period was the wide-spread reforms brought on by the well-known Flexner report (Flexner, 1910). This has been a well-documented source of regulatory influence on all medical education in the United States (Ludmerer, 1985). State governments systematically responded to Flexner's call and made sweeping changes to state licensure processes for medical doctors and for medical training facilities. LLU was big enough, at least to its SDA church, to believe it should and could meet these reforms. However, LLU was small enough that it would have to focus all of its formative identity development on this task if it was to succeed. Accepting such challenges for change have a profound way of shaping identity. It is these hard tasks that define character. That is precisely what happened for LLU during this second formative period.

LLU's intense desire to be accredited led to a blanket responsiveness to the AMA. During this period, this AMA allegiance was firmly set and played strongly into LLU's later development. Alliance with the medical establishment for accreditation provided a bridge for AMA ideologies, values and practices to be . transferred to LLU. LLU's rush "to comply" with the AMA is understandable. However, from this period on it becomes unclear to what extent this allegiance forces LLU to sacrifice its other value commitments and therefore negatively distort its own uniqueness. As I discuss later in this chapter, organizational survival itself is important for institutional integrity. A dead organization can support values much like martyrs can support their causes, but dead organizations can not continue to act on those values.

Despite this apparent across the board allegiance to the AMA during this and later periods, one event from this time indicates that LLU still had religious values it would refuse to sacrifice. At the start of World War I, LLU was still an AMA "C" rated school, seeking to move to the higher "B" rating and eventually to an "A" rating. A "B" rating was needed by all medical schools if student's in those schools were to avoid the draft. A draft would have drained LLU of students at this time. Fortunately, not long into the War, LLU secured the "B" rating in 1918 because of the construction of a new acute care hospital in Los Angeles (the Ellen G. White Memorial Hospital) (Neff, 1964). But that was not sufficient. "In the summer of 1918 the military authorities issued new regulations requiring that in order for medical students to be deferred, they must be enlisted in the so-called Student Army Training Corps" (Shryock, 1984, p. 102). However, the idea of a Student Army Training Corps disturbed some SDA leaders who wanted nothing to do with training for war. As Shryock noted,

> The church's definition of noncombatency forbade the carrying of arms but permitted such participation in military activities as could be performed otherwise. The role in which an Adventist man in uniform could fit best was in the care of the sick and wounded. The reason for objecting to membership in SATC units was that this involved voluntary enlistment in which, it was assumed, the individual signed away his right to noncombatant status. (p. 102)

LLU looked for alternatives to preserve its own values, keep its SDA leadership content, and still respond to coercive regulatory demands. One option was to have "medical students of Loma Linda Campus ... perform their military drills with students at the Redlands University, and students in Los Angeles could become part of the SATC at Occidental College" (p. 102).

In addition to the issue of noncombatency and participation in SATC, the existence of a two campus medical program was also cited as reason not to exempt LLU students from the draft.

> This time Dr. Evans [LLU President] traveled to Washington. But his efforts were fruitless. Five CME students had received summons to appear for induction. The day for them to be inducted soon arrived. While they were awaiting transportation to the induction center, their orders were suddenly canceled. It was November 11. The armistice had been signed and the war was over. Once again Providence had intervened, events were timed just right, and the School of Medicine survived. (p. 102)

Neff (1964), argued that "only divine intervention prevented the collapse of the medical college—a collapse which would have ended the denominational medical-education program for all time!" (p. 208).

Despite the fact that these historians dramatized the potential negative outcome of these events, this story shows that identity formation occurs with each new crisis. Choices are made. Directions are set for later development. LLU survived this event slightly altered and compromised, but still clinging to its dual allegiances. It was committed to and responded to (1) a sense of divine calling to train SDAs to minister the gospel to the physical and spiritual needs of others as well as (2) the need to adhere to federal and AMA regulations. For LLU, the showdown between its spiritual values and external demands was attenuated short of life threatening outcome or a bitter internal civil war. A "near death" experience like this one can have a profound influence on organizational identity development. For example, a standing Army hospital was created at LLU as a result of this experience. That hospital was later fully activated in World War II. Meeting AMA regulations were viewed as a great saving force during this period. As a result, meeting AMA demands became even more acceptable to LLU leaders after this event. Finally, being "saved" from this "near death" experience reconfirmed for many participants LLU's divine calling and further solidified SDA commitment to LLU.

By 1922, LLU received an "A" ranking from the AMA as a direct result of its construction and utilization of White Memorial Hospital in Los Angeles. That hospital gave LLU increased involvement in acute care in the Los Angeles County area. It put LLU faculty and students into close contact with the medical needs of those in its region. Around this time, the University of Southern California's medical school closed for almost a decade, during most of 1920s. This gave LLU even more commitment to acute care. LLU became a central player in the building and staffing of the 1933 Los Angeles County Hospital. All this occurred at a time that the county grew from 936,000 to 2,208,000 (135%) in a decade.

In summary, national defense issues nearly forced LLU to shut down but its active response to accreditation demands gave it an acute care facility and identity at just the right time. This acute care identity would grow quickly. Closure of the USC medical school, the needs of a growing county and the construction of a new county hospital would guide LLU into a new direction of work. The "city" and "acute-care" identity would slowly start to overshadow the more rural, natural remedy qualities at work at the Loma Linda campus. From this period on, LLU would never lose its need and its identity as an acute care powerhouse. After the 1930s, new medical colleges and top notch hospitals poured into Southern California and displaced LLU's major role in Los Angeles. However, even when LLU united its training facilities back to Loma Linda (60 miles inland) in the 1960s, the acute care presence of the White Memorial Hospital remained.

3.3 Early Phase: Period Three: The Stable Period, 1930-early 1940s

In 1930, LLU's 25th anniversary created interesting identity discourse. In 1930, the medical director (Walton, 1930) at Loma Linda University's Loma Linda Sanitarium hospital wrote:

> The purpose, ideals, and spirit which actuated the founding of this institution have been handed down to us as a sacred heritage to be cherished and safeguarded.
>
> The conception of the idea of ministry for others, and the inspiration and example for such a benevolent work as this centers in Christ, the Great Physician....
>
> Loma Linda is not just another medical institution. Its work has been and is ever to be unique. Loma Linda is an educational center, a center for the training of skilled physicians, nurses and other medical workers who are to minister to the world's need.... where the sick can obtain the most competent medical and surgical care at the hands of skilled physicians and nurses; a place where patients are taught and impressed to regard and obey the laws of healthful living...a place where the "law of the Spirit of life in Christ Jesus" is made plain.
>
> Our hope and determined purpose for the future is that Loma Linda shall in even a greater degree fulfill all the design and original purpose for which it was founded; that is shall be indeed a city of refuge where the weary and sick of earth may find and experience spiritual refreshing and healing of bodily ills. (p. 1, 2)

Looking for one's uniqueness is not only a fundamental human longing, but it is also an important task as administrators seek to garner public support and legitimacy. However, LLU leaders seemed ready to balance the need to be unique and special with the need to distance themselves from those who would be considered religiously weird. In 1931, LLU's president, and an accomplished orator, P.T. Magan was invited to explain LLU's religious identity and its unique approach to medical education at the prestigious Association of American Medical Colleges. His address (Magan, 1932) was also an identity statement about LLU. He felt medical students should "be impregnated with ... divine characteristics" and

> the most important qualifications the doctor should possess are not knowledge of medicine at all [but] character, sympathy, industry, charity, patience, economics, the spirit of consecrated service—in a word, the art of medicine, as it was once understood—medicine as inculcated by the Master.... It was more the human qualities of the immortal Osler than his scientific attainments that endeared him to his disciples and patients. Fortunately, he was super-endowed with both knowledge of the humanities and of the sciences. (p. 3)

Magan used strongly religious tones but couched LLU's unique ideologies in a wider medical ethic. However, to insiders, his comments were in keeping with SDA dogmas. Nevertheless, Magan was eager to distance his organization from other religious groups that claimed to blend religion and medical science.

> We wrestle not against innocent forms of harmless quackery but against gigantic systems of cultist evils, against spiritual wickedness in high places—Mary Baker Eddy— that Lydia Pinkham of the soul, and Amy Semple McPherson, the prima donna of medico-religious vaudeville. (p. 2)

At times some LLU leaders may have tried to legitimate the organization to "outsiders" in such a way that some felt its identity was compromised. However, LLU balanced its two identities of religious service and AMA accredited medical education, producing students for world missions and regional service, as well as for traditional U.S. medical work. However, global changes related to World War II (WW II) brought pressures that would lessen the need for medical missionaries, and refocus LLU toward a regional educational facility.

3.4 Late Phase: Period Four: The Adaptive Period, mid 1940s-mid 1960s

From WW II to the mid 60s when the name of the institution was changed from the College of Medical Evangelists to Loma Linda University, LLU experienced rapid change. In this adaptive period, changes in California demographics, post-secondary education accreditation and structure, federal involvement in health care, and the influx of baby-boomers to college, all changed LLU. While this period altered LLU identity in profound ways, I only have space to briefly report two significant changes— the consolidation to Loma Linda and the move to a comprehensive university structure.

Since the mid 1910s, LLU had a two campus program. It had nursing programs at both its Loma Linda and Los Angeles campus and medical students divided their time between both campuses, the first two years at Loma Linda for basic science training and the last two years and internships primarily in Los Angeles. Repeatedly LLU received internal and external pressure to merge the two campuses, but consensus was never reached. By 1960, LLU's board made repeated property purchases and building decisions related to campus consolidation and repeatedly revoked or let these decisions idle because of internal disagreements or external pressures. This usually ended up with each campus (Los Angeles and Loma Linda) growing a little on its own and becoming even stronger as a potential site for consolidation.

In 1960, a committee was set up to give a final report reviewing all the facts and the conflicting views about consolidation (*Board Minutes*, February 8, 1960, p. 116). It was felt at this time that a decision had to be made. Disgruntled faculty, especially those in the clinical division (Los Angeles), had communicated directly to the AMA about their desire for consolidation to Los Angeles (*Board Minutes*, April 6, 1960, p. 141). The Board was upset by their actions, but this motivated the Board to invite AMA consultants to advise the college about consolidation (*Board Minutes*, April 6, 1960, p. 141). Tensions were high during this period. A lot was riding on this decision, not just because of school buildings but also because of faculty homes and medical practices/offices (personal communication, Harold Shryock and Marilyn Crane, 1998).

The AMA consultant's report was sensitive to both sides of the argument, but supported consolidation to Los Angeles. However, their listings of the pros and cons for consolidation to either campuses fueled even more indecision in LLU leadership (*Board Minutes*, September 12-15, 1960, pp. 179-191). No decision resulted. However, the minutes of that meeting (*Board Minutes*, September 12-15, 1960, pp. 179-191) and the report generated for that meeting (*The Problem of Consolidating the School of Medicine of the College of Medical Evangelists* cited below as *The Problem*.) contain the intricate details that speak to the divergent influence of founding identity on later identity-shaping decisions. Each side was not only quick to give the technical benefits of their view, but each was also eager to claim that specific founding values supported its choices. Those pushing for consolidation to Loma Linda noted that "Adventists believed they had divine direction and sanction in the selection of a site for a school at Loma Linda in 1905" (*The Problem*, n.d., part I, p. 2). For them it was "quite clear that Mrs. White acquiesced in the decision to establish limited clinical teaching facilities in Los Angeles; it is equally clear that the messages transmitted through Mrs. White to the Church identified the medical school with the Loma Linda educational center" (p. 3). Other arguments favoring Loma Linda centered around the value of close faculty-student relationships, the establishment of a distinct identity, locating schools outside of the city, and having full control of a hospital for purposes of spiritual witness (pages of Part II and rebuttal that seems to be Part III).

The group supporting the consolidation of the school to Los Angeles also made their appeal to founding values. They quoted an early LLU President and White's statements about the need for healing ministry in the city. Los Angeles gave LLU "an opportunity to educate medical evangelists on the plan outlined by the Spirit of Prophecy [Adventist informal language for Ellen White]" (Part III, p. 2). They went on to argue that very little character changes occurred after 17 and that concern about the city's bad influence on students was overplayed (p. 4). Interestingly, the AMA consultants had made a similar argument but the age was 21 (final section of report, p. 8). The faculty also noted that the environment of Los Angeles prepares youth for mission service because LLU had a greater variety of ethnic groups due to immigration to the area (Part III, pp. 2-4). They also noted that White

> was always more concerned about the broad plans for the medical work than she was about the geography, important as that might be. It is evident in her writings that her plans were broad enough to include Los Angeles, and that she wanted the medical school to continue to do its best. To establish the medical school in any situation where the medical work would be harmed would be against the Spirit of Prophecy counsel. (p. 5)

The pro-LA group also used the Battle Creek experience of the 1890s as a reminder of another important issue. They noted White's warnings about the "dangers of concentrating" too many Adventists operations into one area and reminded the readers how she ceased to support Battle Creek when it got too large, even larger than the Loma Linda campus had become at that time (p. 5). At one point they raised the moral claim "The Seventh-day Adventist church has a moral responsibility to continue the work in Los Angeles" (p. 3).

Conflict over the consolidation issue demonstrated several important points related to institutional integrity. First, it showed that founding values and decisions often served as a reference for later decisions in an organization. Both sides felt compelled to make their claim using these founding beliefs. Second, it showed that founding values could be interpreted in multiple ways, even to indicate support for opposite decisions. Third, it suggested that founding values could become so ambiguous over time, and be applied so creatively, that it was easy to adapt founding values to support any one of many options. In this particular event, the appeals from both sides must have been so compelling that no decision was reached. Plans for the improvement of both campuses was voted (*Board Minutes*, September 12-15, 1960, pp. 187-191).

The 1960s started out like the 1950s, leadership ambivalence and more starts and stops in facility and program planning. The work of AMA consultants and blue ribbon committees had not pushed past this identity quandary. Nevertheless, fear about a potential loss of AMA accreditation in the next review of 1963 got the Board and administration back to the decision table. Furthermore, contractual arrangements with the Los Angeles County Hospital were scheduled for renewal in 1962. These two forces motivated LLU leaders to consider setting up their own large teaching hospital (*The Problem*, consultant's report, p. 3). At this time, David B. Hinshaw was selected to be medical dean in Los Angeles. While the previous dean, Walter Macpherson, had been a consensus leader, Hinshaw was not so inclined (Johns, 1984, pp. 68-69). Hinshaw brought a stubbornness to the situation that polarized discussions. Furthermore, he was sympathetic to the idea of consolidation to Loma Linda. Given such sympathy by a prominent Los Angeles campus physician, changes were soon to come. Hinshaw believed that having a hospital totally under the control of the university was an attractive option and consolidating to Loma Linda would allow this (*Board Minutes*, September 25-26, 1962 shows pp. 431-432). Another concern was the fear that a move of the basic science program to Los Angeles would result in "at least fifty per cent non-SDA" faculty. It was feared that LLU's distinctiveness would be lost. Given these influences, consolidation to Loma Linda was approved (p. 439)

By late 1962, the decision to consolidate to Loma Linda focused LLU's energies in a new direction. However, this new resolution was not supported by all constituents. The November 27, 1962 *Board Minutes* reported the cold reception of the AMA to this decision:

> Dr. Wiggins, of the American Medical Association, was not too favorably impressed by the recent action to consolidate on the Loma Linda campus; and implied that there might be an annual inspection with a regular inspection next January. Stress will be laid on what we are currently doing and on what plans we have in the future for upgrading the School of Medicine. (p. 445)

LLU, specifically its School of Medicine, not only had to attend to a closer scrutiny in reference to its long time regulatory agency, the AMA, but it had to do that while it engaged in the difficult process of garnering local and broader SDA support for funding new buildings and the transfer of staff and students to Loma Linda.

In May, 1963, an eleven-story hospital structure was approved and in July, 1967 the new building was occupied as the Loma Linda University Medical Center. This building proved a significant structural center for the emergence of a new identity for LLU, one firmly tied to the Inland Empire region of Southern California.

The AMA consultants had earlier noted that it was "difficult to visualize" an "architectural expression of a unified medical center" at Loma Linda (*The Problem*, Part IV, p. 5). However, the final choice to build in Loma Linda and the resolution this decision produced under Hinshaw gave LLU new energy around an identity that was tied to the geographic center of Loma Linda.

Two other important decisions were made in 1960. The first decision was to involve LLU in liberal arts education. Previously, LLU was exclusively focused on health care programs. A committee was authorized to examine the offering of "three years of humanities in the liberal arts colleges and a fourth year principally of science at the College of Medical Evangelists" (*Board Minutes*, September 12-15, 1960, p. 191). Here the college was opening itself up to a non-health care identity. As this sentiment progressed, merger with the nearby La Sierra College, another SDA college, was explored and eventually authorized. E. E. Cossentine was appointed to head up a task force to examine this issue. Interestingly, E. E. Cossentine was the President of the La Sierra educational facility in the 1930s that recommended that then LLU President P. T. Magan consider the idea of a merger between the two institutions. Thirty years later this merger would become a reality. However, as I explain in the next section, this merger would not last.

The other major decision during the September, 1960 meeting was the change of the college's name. "VOTED to recommend to the constituency an amendment to the Articles of Incorporation, substituting the name Loma Linda University for College of Medical Evangelists" (p. 191). The School of Medicine tells the alumni in a brief note in May, 1961 that

> the name change was easily understood because 1) it denoted plans for a non medical emphasis on campus, 2) alumni clearly favored "Loma Linda University" more than any other name, 3) Adventist leadership supported it, and 4) the other name "has been a handicap in certain respects in gaining admission for practice within certain countries." ("Loma Linda University Name Change," pp. 6-7)

This decision generated controversy. Some were concerned about a loss of a distinctive religious identity. One of the letters I discovered made the basic argument that the university was losing its evangelistic purpose, which was the central distinguishing characteristic of LLU. This particular letter collected quotations from Ellen White and other early college leaders to show that the founders desired a very religious purpose, even a ministerial and a seminary purpose to the school. Those who signed the letter, many of whom were early students or faculty at the school, expressed the belief that the change in name would lead to a loss of unique institutional identity (letter found in General Conference Archives, CME Miscellaneous Historical Papers, R. G #267, Box A 1211). The voice of dissent must have registered with the board who voted "that we seek to preserve and perpetuate the name College of Medical Evangelists as a tradition even though not as an official name" (*Board Minutes*, January 31, 1961, p. 237). Time proved this to be more of a passing comment than a reality.

A corporation's name is important in identifying its commitments, but it can be argued that a name change also provided a significant loss of identity for LLU. The change dropped the unusual "evangelists" from the name. Dropping that "oddity" may have allowed LLU to attract more non-SDA students and faculty. Or, the name change can be interpreted as matching the change already at work in LLU. Regardless of the causal direction, it is self-evident that "evangelist" was a founding value that would be downplayed in the future of the institution.

Oddly, just as consolidation to Loma Linda was a re-establishing the organization original geographical center, the name change to Loma Linda University appeared to hide not only the school's original religious heritage, but also its focus on medicine. The removal of "medical" from the title denoted a move to the then popular idea of a comprehensive university, a change that would not last.

3.5 Late Phase: Period Five: The Expansion Period, mid-1960s – 1990

During the 1960s and 1970s, LLU enjoyed rapid expansion for at least three reasons. First, its name change and its merger with La Sierra College (LSC) to become a comprehensive university attracted a more professional diverse base of students. Second, consolidation to the growing Inland Region of Southern California occurred at a time in which educational markets were growing in that area. Third, it benefited from the general increase in college attendance hitting the nation as a result of baby boomers. This expansion period ends in the 1980s with dropping enrollments and rising tensions between the La Sierra and Loma Linda campuses of Loma Linda University. It is impossible to compress all the growing pains (program proliferation, building purchases, expansion and subsequent contraction) of this period into this chapter. Instead, I focus on the identity influence of the split with La Sierra College (LSC) during this period.

It is difficult to retell the twenty-five year marriage between LLU and the nearby SDA liberal arts college of LSC in Riverside, California without retelling it through the lens of the "divorce" of the late 1980s. As such, I focus mainly on the split as a way of understanding the identity changes that altered LLU during this period and prepared it for the next period of identity formation.

When the regional accreditation organization, the Western Association of Schools and Colleges (WASC), placed LLU on probation in the late 1980s, several Loma Linda campus administrators accused LSC faculty and programs for the problem, and LSC quickly became an easy target for separation. Although the board just before this time had created a special subcommittee assigned to study the dual

campus tensions and suggested consolidation and prepared brochures and plans for
promoting consolidation, WASC probation shipwrecked that plan (see early plans in
Board Minutes, January 11, 1988, p. 21). Concern about WASC fueled interest in
separation (*Board* Minutes, April 20, 1989, p. 28). "A great gulf exists between the
two campuses which must be dealt with." Doubt existed on "whether the campuses
[could] be held together as a 'single university" (p. 27). Five options were
considered "1. Maintain the status quo. 2. Complete separation...3. One Board, one
administration, two accreditations. 4. One Board, two administrations, two
accreditations. 5. Two Boards, two administrations, two accreditations, chancellor
and small staff to coordinate" (p. 31).

The School of Medicine (SM) and Loma Linda University Medical Center
(LLUMC) administrators called for complete separation. One School of Medicine
administrator raised the fear that WASC probation would threaten student and
faculty recruitment at LLU, cut admissions to the LLUMC, and even hurt residency
programs. LLUMC and the medical programs had progressed well since the 1960s.
These fears seemed ungrounded and more for political affect. Regardless, fears by
prominent administrators agitated blame. WASC's probation was tied to the
discontented and bitter La Sierra faculty. The LLUMC administrator argued that
"the School of Medicine was "being 'held hostage' by actions originating on the La
Sierra Campus" and felt that this issue must be "dealt with speedily" and separation
was a quick way to solve what was labeled a major problem (*Board Minutes*, June
22, 1989, p. 40). Many administrators from the Loma Linda area became
sympathetic to this view (p. 42).

At the time, La Sierra administrators acknowledged the shortcomings of their
campus—declining enrollment, reduction in programs, poor plant and landscaping
maintenance, personnel reduction and concern about consolidation— but they
continued to argue that separation was not the remedy (*Board Minutes*, June 22,
1989, p. 44). However, several Loma Linda campus deans persisted in their call for
separation. This was indication enough to some La Sierra faculty that their own
value were being called into question. For them, separation was a way to regain
academic dignity for what they contributed to SDA education in the region. They
also called for divorce.

Lyn Behrens, then Dean of the School of Medicine, probably became the most
persistent voice for separation. She used LLU's famous "Baby Fae case" to further
contrast the growing view of the Loma Linda campus administrators.

> Evidences of rejection have been apparent ever since the Loma Linda/La Sierra merger
> of 1967.... We need to return to the original mission of Loma Linda University as a
> health educational institution. The faculty concerns of the two campuses are different
> and probably can never be fully reconciled. The basic mission is different. Loma Linda
> is to be a place of healing, a place for education in the healing arts, within a fully
> accredited program. It was recommended strongly that the Board disengage from La
> Sierra and set about to fulfill the prophetic mission for the Loma Linda Campus. (p. 47)

With the cadence of a sermon, the confidence of a prophetic purpose, and the
decisiveness typical of some administrators, the call to split was linked to issues of

identity and statements of integrity. This symbolic and administrative pressure swayed LLU President Woods. While originally a supporter of consolidation, he now felt the appeal to "return to the original mission of Loma Linda University as a health educational institution." He now would accept this view, and argue that official statements from WASC supported that decision (p. 49).

Despite another board decision to create "two campuses, two chancellors, and two operating boards" (*Board Minutes*, August 27, 1989, p. 54), the board eventually supported the growing sentiment and voted for a full separation (*Board Minutes*, October 4, 1989, p. 82). While this was occurring, the La Sierra (Riverside) campus had developed a faculty governance structure in response to WASC suggestions to increase faculty involvement in administrative decisions. The LLU board chair and SDA Church President, Neal Wilson, thought this move by La Sierra faculty was unacceptable and "union like." Unions were and still are often deemed offensive by some Adventists. This last straw helped to speed up separation (*Board Minutes*, February 13, 1990, pp. 3-5). "The papers officially divorcing the campuses....were signed in a solemn ceremony on Saturday night, August 25, 1990, in the student chapel of the Loma Linda campus. Board members and many interested faculty attended. Some cheered. Others cried—or at least felt like it" (Cottrell, Walters, Bradley, and Rouse, 1993, p. 13).

The twenty-five year experiment with an identity as a comprehensive university was over. Some saw this period as a step backwards, a wasted opportunity to strengthen a new emerging identity as a comprehensive university with effective faculty leadership. Others saw this as a step backwards but a good one, one that re-established the institution's original and true identity as a health science education organization, with strong centralized decision making. The latter group's vision and view dominated.

Lyn Behrens was chosen as president of the new university (*Board Minutes*, June 5, 1990, p. 56). She had, with Hinshaw, been the main spokesperson for separation and like Hinshaw had a decisiveness that rallied action. She took up the task of refocusing the university back to its health science heritage. She pressed from the board a formal designation of LLU as a health science institution (*Board Minutes*, September 20, 1990, pp. 78-81). A WASC visit in 1991 lead to the removal of probationary status on March 3, 1992.

3.6 Late Phase: Period Six: The Rediscovery Period, 1990-Present

Since the La Sierra split and the successful WASC accreditation in early 1990, LLU has been in a period of rediscovery. I label this period re-discovery for two reasons. First, LLU was now fully re-asserting its health science and religious identity. LLU's President Behrens (1990-present) has had a profound influence on this process. Behrens, an Australian-trained SDA physician who came to Loma Linda in 1966 as a pediatric resident, had worked her way up in the organization (Dwyer, 1992, pp. 5, 9-13). As such she was the first woman Dean of the School of Medicine and first woman president of a large SDA college. "One of the things that particularly endears Lyn Behrens to the board is her spirituality ... Not long after assuming her presidency she reread all of Ellen White's comments about Loma Linda and distributed them to the faculty" (p. 12). My own personal participant observations while a student and faculty member at LLU confirm Dwyer's observations that Behrens has an effective leadership style that focuses on both integrity and religious aspects of LLU.

Nevertheless, the skeptics remained. One influential campus contact wondered out loud to me if Behrens was "using religion" as an administrative tool to garner support. This individual noted the historical precedence of past SDA and LLU leaders to welcome the unifying role of religion without welcoming religion's more radical role in challenging administration and established practices, especially those inherited from the influence of traditional medicine. This person feared that a true community of diversity was not going to be fostered, and cited the absence of administrative humility. It appeared that the American Association of University Professors (AAUP) shared some this person's concerns. AAUP listed Loma Linda University on a short list of institutions that have "censured administrations". AAUP argued that these censored administrations have worked to squelch faculty involvement in governance or fostered poor methods of due process in human resource decisions (See the latest listings of "Censured Administrations", as seen in any recent 2003 *Academe*). Only time will show how religious values continue to work in the life of LLU.

The second point of rediscovery was the attempt to integrate the university and its medical center. In the late 1990s, the LLU and LLUMC Boards created the Loma Linda University Adventist Health Sciences Center (LLUAHSC), which was an umbrella organization to orchestrate the activities of LLUMC, LLU, and several related entities. It is too early to determine the long term impact of the reorganization on LLU's identity and institutional integrity. In 1999 Behrens became the president of LLUAHSC.

The following statement clarifies how the new structure links to issues of historical identity and institutional integrity:

> Loma Linda University Adventist Health Sciences Center, a new corporate structure, has been created to coordinate the academic and health-care activities of our health science institution....
>
> We are accountable for the fulfillment of our mission--to continuing the healing and teaching ministry of Jesus Christ....
>
> We are also accountable as an academic health sciences organization to meet society's expectations for health care and education, which are quality, accessibility, and cost effectiveness...
>
> In addition, Loma Linda has .... the value added of Adventist education.... the integration of faith into learning so graduates can maintain their own wholeness and participate in professional ministry....
>
> From 1905 until 1980, the Loma Linda institutions were one corporation with a single board. Thus, the board was responsible for Adventist education and Adventist health--

ensuring the effectiveness of each and cooperative relationships. [From 1980 the two
institutions were not linked by a common board] ...

Great care has been taken in defining and codifying these organizational changes to
maintain appropriate autonomy of LLU and LLUMC as separate corporations, while at
the same time addressing the need for long-term linkages between them." (*New Loma
Linda Corporate Structure Created*, 1998)

In recent years, it appeared that LLU's re-discovery period gave rise to brisk
discussion of religious commitments and a public focus on "continuing the teaching
and healing ministry of Jesus Christ" (LLU Mission Statement). It remains to be
seen if other changes will also follow as a part of this re-discovery. For example,
will LLU's original emphasis on unique Adventist views of health care and health
reform reassert a stronger place in LLU's educational and health care practices? Will
hiring and admissions practices continue to focus on candidates' individual
commitments to LLU's mission of Christian healing? Will internal resources be
adequately funded and distributed to support SDA religious and health
commitments? How will Christian ministry be integrated into curriculum? Will
administration continue to articulate the organization's corporate call to serve the
health needs of others as Christ did? At what cost will they continue this calling in
the face of the highly competitive health care environment?

This section reviewed the historical identity development of LLU. The next
section uses literature on organizational development to interpret LLU's changes in
order to nuance the complex "institutional" processes sociologists belief shape
organizations.

4. INSTITUTIONS, ORGANIZATIONAL THEORY, AND INSTITUTIONAL DEVELOPMENT

Figure 1 shows the three most identity shaping macro institutional influences that
molded LLU's integrity development. I believe these institutions provided the
cognitive, normative, and regulatory forces that shaped LLU's construction of its
identity. That identity is not merely a wholesale absorption or adoption of
environmental influences, but rather an identity crafted from choices that range from
full adoption to rejection. It fell to various administrators and influential board
members and key staff to determine which ideas settled in and took hold. More
specifically, SDA religious and health care dogma and practices, traditional medical
beliefs and practices, and post-secondary health career educational regulations
shaped LLU's founding values, its unfolding identity, and subsequent self-
understanding of the basic battles it needs to fight to maintain its integrity. I review
some ideas on institutions below and link them back to LLU's development.

The words "institutions" and "institutional" are often used interchangeably with
"organizations" and "organizational" in traditional parlance. However, sociologists
formally use the terms "institutions" and "institutional" to refer to social processes
and pervasive normative, cognitive, and regulatory influences that guide individual
and corporate behavior (Scott, 1995a, 1995b). In general sociology, sociologists

conceptualize macro (large scale), meso (regional or local), and micro (local or interpersonal) institutions as social influences that script our lives as social actors. For example, shaking hands to greet another person is a micro institution common in Western cultures. Many of us naturally do it without even thinking about it because we have been socialized to do so. That is why it is called a "taken for granted" or cognitive or mimetic social institution. Not all institutions are so subtle. Some practices persist because a regulatory agency demands them. For example, accreditation agencies are big at altering university identities by merely the threat of withdrawing accreditation. That is a regulatory or coercive institutional influence. However, professional organizations may also foster wide-spread normative influences in the form of moral demands or codes of conduct.

Micro and meso institutional scripts and expectations operate within larger, pervasive ways of social being called social institutions. Currently social theorists envision at least six macro social institutions. These are: (1) kinship (family relating processes), (2) religion (the way we relate to transcendence, worship, and the mysteries of life), (3) economics (the way we conceive and run our financial and occupational lives), (4) government (our federal-state political processes), (5) legal (our courts and justice processes), and (6) education (the way we train and raise our children). Some sociologists have theorized that both medical science (inquiry, hospital processes, disease,etc.) and recreation (professional and amateur sports, racing, gambling, hobbies, etc.) are developing a critical mass of influence such that they may be the next emerging social institutions in Western societies (Turner, 1997). As such, LLU operates in three specific macro institutional sectors that shape its identity and integrity development—religion, education, and medicine. LLU's loyalties and commitments and its whole identity development and integrity hinge on changes in those macro areas.

Institutional research on organizations utilizes all three levels of institutions with special attention in neoinstitutional theory to macro institutional forces. For them, legitimacy rather than simple or dramatic environmental evolutionary influences explain the mechanisms by which organizations craft themselves. Legitimacy occurs when a person or a group recognizes what it is doing as valued and appropriate to a larger audience. An individual or corporate group comes to envision what is expected of it through interaction with others. These others, these social arrangements, craft the institutions (macro, meso, micro) through which reference points of legitimacy are created.

This legitimacy process is vital to the theories of change offered by institutional theorists. They believe that organizational practices and structures are often less the by-product of "efficiency and market competition" and more the result of attempts by organizations "to create or conform to categories and practices that give classificatory meaning to the social world" (Brint and Karabel, 1991, p. 342) That is why institutional theorists define institutions as the "cognitive, normative, and regulative structures and activities that provide stability and meaning to social behavior" (Scott, 1995a, p. 33). These structures and activities are transformed and transmitted by "cultures, structures, and routines" (p. 33). Applied to organizational

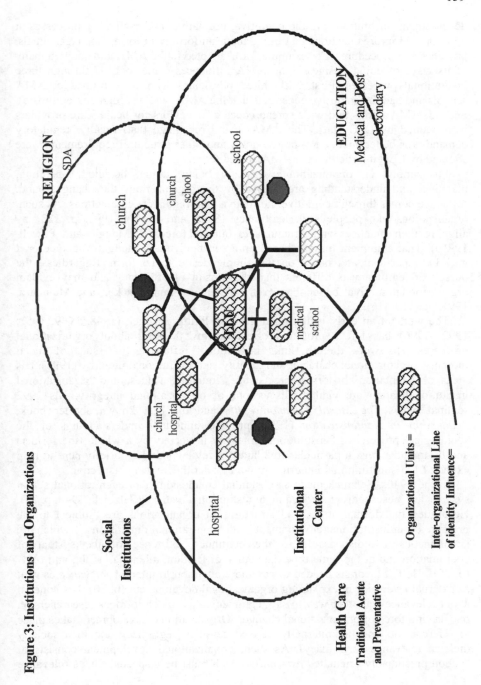

Figure 3: Institutions and Organizations

development, institutions are the cognitive, normative, and regulatory processes at work on and in organizations that bring about uniformity in form, language, rituals, practices and procedures. For example, LLU adapted to the AMA and adopted many of its ways of doing business. The AMA influences came to LLU through three institutional processes. First, LLU hired physicians who were trained at AMA schools and had the cognitive constructs that the AMA would recognize (legitimate). Second, AMA norms and wide-spread codes of behaviors for health care providers were valued by LLU staff. The AMA would recognize that. Finally, regulatory demands for better acute care, more books, and other structural requirements were adopted by LLU in order to gain AMA accreditation.

Put simply, an organization's ways of being, doing, deciding, operating, planning, and understanding are shaped by the organizations they hang around. Influence comes through cognitive, normative, and regulatory interactions. As such, organizations like people seek legitimacy (Scott and Meyer, 1991, p. 123) by aligning their practices and structures to outside forces (DiMaggio and Powell, 1991b). That alignment may be based on a substantive and thoughtful choice or it may be superficial and based on the simple desire to conform. Regardless, the pursuit of legitimacy is different than the pursuit of efficiency and may explain organizational survival better than even efficiency (Singh, Tucker, and Meinhard, 1991, pp. 410-416).

The work of an early institutional theorist, Philip Selznick (1948, 1949, 1957, 1992, 1996) has left a profound effect on my thinking about organizational processes. His work directly links institutional influences to issues of social morality, organizational character and identity, and institutional integrity (1992). His ideas of cooptation, historical precedence, founding values, and organizational institutionalization are vital to discussions of organizational integrity. His 1992 seminal work, *The Moral Commonwealth,* and his best known shorter book, *Leadership in Administration* (1957) provide important understandings of the sociological processes of institutional integrity in society as a whole (the former) and in organizations in particular (the latter). I review his ideas briefly here to help explain LLU's institutional integrity from a broader theoretical perspective.

In the 1940s, Selznick made a substantial conclusion "that even rational action systems are inescapably imbedded in an institutional matrix" (Selznick, 1948, p. 25). He argued that formal and social structures in organizations are "subject to the pressure of an institutional environment" and "non-rational dimensions...[which are simultaneously] indispensable to the continued existence of the system of coordination and at the same time the source of friction, dilemma, doubt, and ruin" (p. 25). He felt theorists needed to pay more attention to internal *relevance* as well as external *relevance* in explaining organizational development (p. 30). His concern about relevance—another way to say legitimacy—did much to focus researchers on one driving force of organizational changes. Organizational development takes place in reference to both an internally derived sense of legitimacy and to a socially dictated relevance (legitimacy). As such, organizational development consists of "action presumably orientated *externally* which must be inspected for its relevance to *internal* conditions" (p. 30, emphasis added). In other words, organizational actors, when looking outward for legitimacy, are at the same time looking inwardly to see if those external realities would better express the meaning of the organization.

This led Selznick to see organizations, like individuals, as forging a "personality" in the presence of social pressures though not in strict uniformity to them but rather in a process of comparing internal processes and past practices to present realities. Organizational mindfulness of its heritage in the presence of new experiences creates a stress on the organization's self-understanding. These dynamic tensions can be avoided for a while but ultimately someone in the organization, typically administrators, must craft meanings of legitimacy in the face of new challenges to organizational identity (p. 31).

Selznick (1948) enumerated possible processes and mechanisms for institutional influence. One of these processes was cooptation, which was "the process of absorbing new elements into the leadership of policy-determining structure of an organization as a means of averting threats to its stability or existence" (p. 34). He saw this process as arising from concerns about legitimacy and survival.

> The significance of cooptation for organizational analysis is not simply that there is a change in or a broadening of leadership, and that this is an adaptive response, but also that *this change is consequential for the character and role of the organization.* Cooptation involves commitment, so that the groups to which adaptation has been made constrain the field of choice available to the organization or leadership in question. The character of the coopted elements will necessarily shape (inhibit and broaden) the modes of action available to the leadership which has won adaptation and security at the price of commitment. The concept of cooptation...is an adaptive response of a cooperative system to a stable need, generating transformations which reflect constraints enforced by the recalcitrant tools of action. (p. 35)

I believe this quote offers a profound statement about institutional integrity, although in complex language. My retelling of this passage is that choices about leadership and about where leaders will come from shape the commitments an organization makes. The earlier choices set longer-lasting commitments. Using LLU as an example, LLU set its course early on by adding a medical emphasis to its evangelistic mission. Then, by adding M.D. presidents and by accepting the constituencies that they brought with them, that medical focus strengthened into a core identity. When organizations, through leadership actions or policies, make commitments to (a) a specific external group, (b) a certain form of organization, (c) other organizations or institutions, or (d) certain types of goals, they are making commitments that constrain and mold their identity development for years to come. This is cooptation. The results of this cooptation are not just structural but deeply influence an organization's self-understanding and what it comes to value as legitimate.

LLU was formed by its early commitments to the SDA community, specifically that small group within the SDA community that saw health evangelism as central. That commitment to health evangelism linked LLU to the emerging world of medicine. That world changed drastically from 1905 to the present and LLU has had to change to keep pace. Early identity choices can be undone, but more often than not they are merely revised to meet changing demands. As such, the organization changes. That change can be subtle on participants because they are imperceptibly influenced by the cognitive, normative, and regulatory influences that shape them and their organization.

Later Selznick (1949) enumerated other processes of institutional influence at work in a specific organization, the Tennessee Valley Authority (TVA). First, he noted that rationality and concern about efficiency were insufficient to explain organizational development. Organizations need money to survive and efficiency preserves that, but they also need meaning to gain stakeholder support. Workers need a reason to work. Supporters need a reason to give. Simply being efficient doesn't explain the driving force of action. When LLU decided to consolidate its Los Angeles campus to Loma Linda, issues of efficiency were present. Issues of control were also important. However, struggles over identity became the main arguments by which the organization crafted a meaning from external pressure to change. The result was to locate to Loma Linda. Institutional, catastrophic, or contingent influences continually bombard organizations, forcing them to deal with issues tangential to their core operations. New directions are proposed and those are debated within the framework of past commitments. Those debates are themselves identity-forming.

A second issue Selznick noted was that "informal structure" can be stable or institutionalized within an organization so that organizational stakeholders "control the conditions of their existence" (1949, p. 251). Nobody likes uncertainty. Organizational environments always add uncertainty. Organizations, to give some continuity to the environment and to their organization see sources of stable meaning. Organizational members then imbue their experiences with meaning beyond the technical requirements of their work. This is institutionalization. This is the substance of organization identity. This is the fundamental reference point for institutional integrity. For LLU, that identity was and to a great degree is still medical evangelism. While the names of "medical" and "evangelism" were dropped from the university's name in 1961, connection between faith and health have been central to LLU's development. LLU has maintained that link in its educational programs and in its mission statements even though it has changed what it calls this central mission and how it explains that central religo-medical mission to its contemporaries (Covrig, 2001b).

Third, mechanisms of leadership, communication, and persuasion develop to fill the space between external forces and internal self-understandings (institutionalized forms). This is the public relations and visionary leadership aspects of an organization that bridges internal self-understandings with external points of legitimacy. This is the place where strong self-statements are needed by central leaders to prevent loss of organizational identity and thus weaken the organization's claim on stakeholder interests. This is also the place for compromise. Both were evident when LLU commitment of non-combatency was threatened by WWI requirements that LLU operate student military operations on campus. At first unwilling to compromise, LLU later formed a standing military hospital. The hospital could claim legitimacy to its own value of "continuing Christ's healing ministry" while still showing its dedication as an "army" which legitimated the organization to the government which valued national support in military protection and preservation. Leadership, communication, and administrative persuasion were the mechanisms by which identity was both preserved and altered in response to internal values (institutionalized identity) and external demands (broad social sources of legitimacy).

Finally, throughout these processes, "human structures" and "commitments to others" will be central factors in constraining organizational identity (Selznick, 1949, pp. 252, 253). These commitments show up in the "uniquely organizational imperatives", "the social character of the personnel," and the nature of institutionalization or the "value-impregnated status" of the organization in its environment (p. 256). All these processes are evident in LLU' struggle to hire faculty and staff who share its value, an increasingly difficult task as the organization grows and seeks to find SDAs capable of sharing LLU's SDA medical evangelist passion to "continue Christ's healing and teaching ministry."

Selznick was able later to simplify his complex understanding of these processes in his influential book *Leadership in Administration* (1957). Using the metaphor of "personality" development, he argued that early influences and choices in organizations influence identity development. Once the organization's structure, mission and ethos were set, further adaptations would have to be made in reference to these early commitments. By this process organizations developed "character" (read identity) which fostered stability (read integrity) even in the presence of turbulent change. Leaders, he argued, had to re-articulate the organization's character with each new internal or external influence. For example, P.T. Magan's 1931 presentation to the AAMC was an attempt to restate LLU's religious based medical education to an increasingly secularized industry. In doing so, Magan distanced LLU from other religio-medical experiments like the Christian Science Mary Baker Eddy and other religio-medical groups that didn't utilize AMA methods as LLU did.

Selznick's metaphor of "character" or "personality" provided a useful way of thinking about institutional identity. It was a metaphor that suggested change was as crucial as resistance to change in understanding integrity. "The formation of an institution is marked by the making of value commitments, that is, choices which fix the assumptions of policymakers as to the nature of the enterprise—its distinctive aims, methods, and role in the community.... The values in question are actually built into the social structure" (pp. 55, 56). Once an organization was established, "Self-maintenance becomes more than bare organizational survival; it becomes a struggle to preserve the uniqueness of the group in the face of new problems and altered circumstances" (p. 21). In this context, the goal of a leader was to "specify and recast the general aims of [the] organization so as to adapt them, without serious corruption, to the requirements of institutional survival" (p. 66). LLU President Behrens appeared to understand this as she pressed for renewed spiritual emphasis in

LLU during her presidency. Her actions to support religious discourse and structural units in LLU indicate a fundamental belief that religious purpose was the value added aspect that differentiated LLU health care education from other health care training facilities.

To this end, two extremes were to be avoided—opportunism and utopianism. The former sacrifices fundamental organizational identity to external pressures. The latter fosters an idealism that prevents the organization from meeting the real needs of the world around it. Utopianism also suggests a situation in which an organization refuses to adapt and improve its own self-understanding as more awareness of its uniqueness becomes available. "If opportunism goes too far in accepting the dictates of a 'reality principle,' utopianism hopes to avoid hard choices by a flight to abstractions. This too results in irresponsibility, in escape from the true functions of leadership" and from the tough task of defining organizational identity in response to specific changes in a turbulent society. On one end of the continuum the organization is too flexible in defining their purposes and each new fad is made to alter that identity. On the other end of the continuum the organization is too rigid in its response to change and thus eventually dies or worse, lives but becomes obsolete, leeching resources from other more useful organizations (pp. 143-149). Change that preserves and even improves character is part of "normal" maturation and fundamental to the growth of integrity.

Institutional theory of organizations has generated a lot of creative ideas and research since the 1950s. Space does not allow me to detail Selznick's later work on blending sociological and ethical insight, let alone the insightful contributions of neoinstitutional theorists like Meyer and Rowan (1991) and others (Powell and DiMaggio, 1991a, 1991b). Neoinstitutionalists have shown how myth and ceremony in the environment get incorporated into the fabric of an organization. These theorists emphasized the role of cognition and language, beliefs, procedures, and roles in organizational development, and they showed how "symbolic" processes influenced organizations even more than technical issues (Scott and Meyer, 1991).

The contributions of institutional theorists continue to provide empirical as well as theoretical explanations of institutional integrity under environmental forces of change. Specifically in healthcare, no other single source is more useful than the well-researched *Institutional Change and Healthcare Organizations* (Scott, Ruef, Mendel, and Caronna, 2000). That book details the changes that influenced hospitals in the San Francisco Bay Area. Their study revolved around two contradictory discoveries: First, both their study and other micro (case study) and macro (population-wide) studies showed that organizations were not only influenced by their institutional environments but by their own actions and that political arrangements can influence whole institutional sectors (p. xix). Second, they noted that healthcare demonstrated signs of profound change and stability. "The changes in medical service delivery that have occurred within the lifetime of adult Americans are truly remarkable" yet many "structures and practices [remain] buttressed by the power of professionals and community leaders, they do not easily adapt to new ideas and new ways. Change is not instantaneous in social systems: these structures exhibit much – perhaps, desirable – recalcitrance and inertia" (p. 1). However, their research documented how change occurred within the broader institutional processes of health. They document how this change varies across the population of organizations and varies across levels or areas within an organization. Their findings imply,

> as institutional scholars have argued, that features of organizations that are not easily buffered from external change (e.g., goals, size, linkages) will be responsive to environmental contingencies. The technical core, on the other hand, which is ostensibly sealed off from most such disturbances ...will display a more constant level of structural inertia. (p.109)

This section reviewed select sociological understandings of institutions and organization change that would apply to integrity issues. A key point was that *institutions* act as pervasive cognitive, normative, and regulatory influences on organization's self-understanding, identity development, and integrity. Institutions influence from both the outside and inside. This is not a passive process. It is a negotiated process, a contested process. Values and sources of legitimation are selected by organizational participants and leaders, and those selection processes are criticized by stakeholders. Each new environmental influence tempts toward opportunism or isolationism/utopianism. Leaders play the role of responding to change and re-articulating identity once change occurs.

4. LLU FINDINGS AND DISCUSSION

In review, I have documented some key points in LLU identity development through six periods of development and used sociological theory to explain that change. First, in the founding period, 1900 to 1915, LLU built its founding identity from conflicts and beliefs operating within its SDA community and took advantage of the health enthusiasm and migration the led to Westward expansion. LLU tapped into the growing SDA passion for doing health work as religious work. In the establishment period, closure of the new school was avoided in the tumult of the World War I draft, and an "A" rating was secured that gave LLU a critical foothold in Los Angeles acute care medicine. This period of growth changes its identity to a leading acute care facility in the region, an influence it still maintains to some degree in Southern California, even though the region now boasts five major medical training (MD or DO) universities. From 1930 to mid-1940s, LLU successfully balances its religious and medical identities into a stable period of growth in students, programs, alumni base, reputation, and fund-raising.

Massive global political changes related to World War II decreased the possibilities for world missions and brought increased drug use in health care. LLU entered an adaptive period as it responded to increased regulation and federal involvement in health care, changes in medicine brought on by increased medical research and pharmacology, regional population shifts, and increased secularization. This adaptation period ended in a name change and a consolidation with La Sierra College as the College of Medical Evangelists became Loma Linda University, a comprehensive university. These changes allowed LLU to prosper from the surge of baby boomers going to college and to enjoy a heavy growth in enrollment and influence in this expansion period. By the late 1980s decreased enrollments, faculty and financial problems, and related issues motivated WASC to put LLU on probation. This brought a certain degree of soul-searching and a fair amount of finger-pointing which fueled already high La Sierra/Loma Linda tensions and resulted in a campus split. LLU has since reaffirmed its emphasis on the health sciences and re-articulated its unique contribution to the SDA Church, the region, and medical education in general. Lyn Behryns' leadership has personified this re-discovering period as she has re-focused on the spiritual and medical identites of LLU and steered the organization away from its comprehensive university status. It is still unclear when the re-discovery period will be complete and what it will produce.

For some, LLU's founding identity, and hence its integrity have been abandoned. They can point to LLU's constant willingness to adapt its original evangelistic and medical sectarianism to new medical and educational practices.

For others, LLU's identity has matured and LLU has become a better, more integral, more purposeful, and more stable organization because of its multiple commitments. Its adaptation has been healthy and part of normal development and maturation process that results from healthy response to macro, meso, and micro institutional influences. For them, LLU has kept pace with the changes of technological and environmental influences. For them, LLU has remained a prophetic organization for a prophetic people. They would argue prophetic people are continually ready to change and to move to new understandings and to "follow the Lamb whithersoever He goeth" (Revelations 14:4) (personal communication with several SDA religious leaders).

For still others, LLU clings to religious traditions and health ideologies that make it unwilling to lose its religious heritage and move on to becoming a top-notch research university dedicated to real advances in the medical science and medical education. They would like to see LLU lose its "weirdness" and become like the rest of the medical community. LLU leaders will have to continue to listen to these many perspectives and articulate their organization's *raison d'etre*. They will, as their predecessors, have to mix old and new terms to explain and thereby craft the organization's ongoing identity.

Regardless of these three viewpoints, it is evident that religion was, and still is, a sustaining influence on LLU's identity. That religion has not been an anchor holding LLU to a specific point in history nor to a certain set of activities. It has been a guiding presence in LLU's ongoing re-definition of itself when faced with new challenges. That religious identity has been and continues to be tempered by other commitments. *As such, holding commitments to multiple communities has become the main institutional engine that forges LLU's identity and shapes its integrity.*

Making multiple commitments kept LLU from drifting in many possible directions. Its religious commitments kept it from becoming just another health care training facility. Its educational commitments kept LLU involved in inquiry, research, and educational ideology. Not all early SDA health care facilities preserved that educational identity and not all have responded to the changes that free and open educational inquiry foster in a religious based organization. LLU did.

LLU's commitment to medicine and research kept LLU from becoming an "other-worldly" enterprise dedicated to sectarian oddities and unfounded and ungrounded medical and religious ideologies. It also kept LLU linked to others in demanding Christian service. Through health care, LLU was connected to the physical, psychological, and social suffering of others. That has given LLU a pathos and a lively experience of serving as Christ would serve. Entering into acute care worked to give LLU a clearer sense of the demands of dynamic Christian service. It helped to prevent utopianism and idealism from disturbing LLU's identity. Nevertheless, acute care and alignment with medical establishments have also brought a dilution of missions. Medicine and health care brought greater wealth and prosperity and made scientific skepticism a value among a community that was built around religious faith. As scientific methods took deep root, religious beliefs and practices were overshadowed. This effected in some older employees a sense of loss. It also created internal conflicts in dealing with issues of faith healing and alternative natural therapies.

In his analysis of institutionalization and the SDA church, SDA professor and historian George Knight (1995) argues that Adventists survived because they were religious but willing to adapt their religious idealism to the challenges they faced. His comments could apply to LLU. He noted that some types of institutions die because they "live in the past as if the past can somehow be preserved intact in perpetuity as a golden age" (p. 158). Other groups "focus exclusively or almost exclusively on the future" (p. 158) and lose sight of the identity forming role of the past and the realities of the present. Other groups, in hot pursuit of relevance, focus exclusively on the present and craft an identity that "has lost its metaphysical foundation [and] its biblical roots in the supernatural" (p. 158). "Relevance without sufficient rootage ends up being irrelevant in the long run" (p. 158). He argued that the best solution was found in organizations that set "forth a present orientation in the framework of the continuum of the past and the future" (p. 158). They fostered "a cosmic viewpoint that finds identity for the present in both history and prophesy. Thus its relevance, being rooted in the great continuum of history and change, is not transitory" (p. 158). This seems to characterize LLU's struggle.

LLU's future commitment to itself and its identity depends on its ability to meet a caution and a challenge that arises from its own faith tradition—a bible passage and an interpretation provided by SDA pioneer and "prophet" Ellen White. The Bible passage relates to King Solomon's leadership of Israel and his country's relationship with other nations: "You shall not mingle with them, neither shall they mingle with you, for surely they will turn away your hearts after their gods" (I Kings 11:2, Amplified). White's interpretation is directed at SDA organizational leaders:

> [Solomon] reasoned that political and commercial alliances with the surrounding
> nations would bring them to a knowledge of the true God; and so he entered into

> unholy alliance with nation after nation..... Heathen customs were introduced....The Hebrew faith were fast becoming a mixture of confused ideas.
>
> The Lord desires his servants to preserve their holy and peculiar character....Too often they themselves, entrapped and overcome, yield their sacred faith, sacrifice principle, and separate themselves from God. One false step leads to another, till at last they place themselves where they can not hope to break the chains that bind them.
>
> Those who are placed in charge of the Lord's institutions... shall not walk contrary to the sacred principles of the truth....God calls for men whose hearts are as true as steel, and who will stand steadfast in integrity, undaunted by circumstances....He calls for men who will not dare to resort to the arm of flesh by entering into partnership with worldlings in order to secure means for advancing his work—even for the building of institutions.... Men to-day are no wiser than [Solomon], and they are as prone to yield to the influences that caused his downfall... God's people to-day are to keep themselves distinct and separate from the world, its spirit, and its influences.
>
> There is earnest work before each one of us....In all our institutions, our publishing houses and colleges and sanitariums, pure and holy principles must take root. If our institutions are what God designs they should be, those connected with them will not pattern after worldly institutions....They will not come into harmony with the principles of the world in order to gain patronage....God would have us learn the solemn lesson that we are *working out our own destiny*.... God is fully able to keep us in the world, but not of the world...Ever he watches over his children with a care that is measureless and everlasting. But he requires us to give him our undivided allegiance.... (White, 1906)

This passage and White's commentary present a sobering caution to LLU leaders. When making multiple commitments, they re-make us. Nevertheless, isolationism is not a cure. Christ Himself argued through the words of a parable that "the children of this world are in their generation wiser than the children of light" (Luke 16:8). As such, some of the most identity preserving reforms may actually come from the children of the world. "External influences" outside the church and church-owned institutions may actually provide more noble and better and wiser influences on LLU that those fostered by its church. Changes to meet these external influences may improve LLU's religious mission. Figuring out which influences will help and which will dilute the mission is the tough stuff of leadership.

5. CONCLUDING SUGGESTIONS

Several ideas from this study may be applicable to other healthcare institutions who struggle with religious identity and integrity. First, founding values and the forces that led to them should be studied to understand institutional integrity. Second, organizations, like people, are products of multiple communities. Questions about commitments and integrity ultimately lead to questions about "commitment to whom and for what." When caught in multiple communities, organizations can be selective to some extent on which aspects they adopt to remain legitimate and which they let go because they are no longer legitimate.

Third, living in multiple communities requires border-crossing. Border-crossing requires juggling multiple meanings simultaneously and articulating organizational definitions to varied constituents. Rare will be the leader that can perfectly balance those commitments. As such, each new leader will naturally be on one side of the border more than on the other. Boards need to realize that when they hire central administrators.

Fourth, the very challenge of border-crossing can be itself a way to preserve integrity. Ethical values from one community can be used to curb the abuses of another. As I have noted elsewhere with regard to professional relationships, maintaining multiple commitments may be the very engine that keeps moral sensitivity alive and integrity a point of thought and action (Covrig, 2001a). The same may be true for organizations. Maintaining multiple commitments and engaging in the work of border crossing may be an engine that forces an organization to make tough choices about its identity and integrity. This can greatly enhance institutional integrity. Identity is formed in conflicts. Integrity is enhanced as values from one community can be used to critique those of another.

Fifth, in times of uncertainty, in a world of border-crossing, and in a world of with multiple communities, organizations with religious commitments may exhibit high levels of anxiety that lead to a perennial need to make self-identity an issue of debate.

> People do not live by reason alone. They cannot calculate and act rationally in pursuit of their self-interest until they define their self. Interest politics presupposes identity. In times of rapid social change established identities dissolve, the self must be redefined, and new identities created. For people facing the need to determine Who am I? Where do I belong? Religion provides compelling answers.... In this process, people rediscover or create new historical identities. (Huntington, 1997, p. 97)

Finally, identity without conflict remains an untested and an underdeveloped identity. Identity forged from and in conflict leads to a stronger integrity. As such, commitments need not be overlapping or congruous to have a useful impact on an organization's integrity. In fact, the presence of varying or even conflicting commitments may create a tension that keeps dialogue and debate about purpose alive. Tension forces administrators toward a deliberateness that requires them to continually monitor internal and external processes. This can have a great impact on organizational readiness for uncertainty and longevity (Weick and Sutcliffe, 2001). Administrators forced to continuously hear multiple voices and painfully select among competing options are more likely to know how their organization differs from others and articulate that uniqueness as a call to integrity. This may help administrators resist institutional forces—mandated uniformity (regulatory accreditation), subtle similarities (institutionalism), and diluted but shared values (institutional norms)—that push organizations to uniformity. I can think of no more painful process by which an individual organization can attend to its own integrity.

Larry May's (1996) observation in the *The Responsive Self* could be made of the responsible organization

> The crucial dimension of maturation for integrity occurs when the self is able to make adjustments in its personality: to block certain influences and to add new ones, thereby beginning the task of providing a unifying structures to the self that is in keeping with how one wants to be. The person of integrity does indeed exercise control over his or her life, but the way in which this occurs is quite different from that proposed by those

who postulate the split-level conception of the self. On my view, both the self that is integrated and the self that acts to achieve the integration are products of factors (such as the influences of family, teachers, ministers, friends, partners, etc.) that are at least initially outside the control of the self. Maturation is largely a matter of 'bootstrapping'—that is, learning to control certain aspects of the self by using other aspects of the self that were initially beyond one's control. (p. 17)

Despite these normal conflicts of identity development, one must acknowledge that environmental forces and internal turmoil may reach a crisis level that can overwhelm an organization. External demands and unanticipated changes and internal dissension about purpose and direction can become so challenging that an administration is unable to hold the organization together. In such situations, organizations face many possible outcomes. These outcomes will vary based on the size and age of the organization as well as by the extent and nature of the external or internal conflict. Young and small organizations may be vulnerable to smaller chaotic events. They may also be more idealistic and willing to die rather than change. As such, *organizational martyrdom* may be viewed as an extreme case of institutional integrity at a primitive level of operation. However, martyrdom among organizations would be a difficult concept to understand without more extensive search.

Second, even medium sized organizations can experience severe instability in their region or their sponsoring organization such that they can be wiped out. An example helps here. A close friend was President of the SDA College in Rawanda. The ethnic strife in Rawanda closed down the school. The college had served five French-speaking African countries and the growing SDA community in that region for decades. SDA higher education is currently struggling to rebuild in that region. Because of political instability, regional leaders have decided to start several smaller colleges in more regions so that all its work will not be impacted if a college in one area is shut down. Catastrophic change can kill an organization. It is impossible to speak about identity if an organization is dead.

Third, diversification, by place or by practice, increases organizational survival even as it can dilute organizational identity. But diversification is often what organizations do as they grow. That is how they withstand pockets of uncertainty. As an organization ages and diversifies (two separate but often linked phenomena), the organization typically develops diverse value commitments and institutional alliances such that more creative ways of compromise are envisioned. In short, it appears that the longer an organization exists, the more a desire for organizational survival breeds a skill at compromise. As more and more people have vested interests in the survival of an organization, the desire for compromise gains momentum (i.e., stakeholders would rather have at least a weakened organization than no organization). In such a way, an organization becomes itself a value that needs to be protected and nurtured. Survival becomes synonymous with integrity.

Fourth, another possibility is that an organization may actually succeed in its founding mission and cease to have the need to exist. Organizational leaders may actually see that their founding purposes are fulfilled by the larger society and that they need no longer exist. As such they can close shop, sell-out to another organization, or adapt their practices and mission to fulfill a whole new set of identity processes. This may be a basic viewpoint among some religious or community based hospitals to sell out to corporate or government agencies (based on conversations with a former administrator of a Catholic institution and local civic leaders).

From my historical analysis of LLU and from my review of institutional theory, I view organizational integrity as

1) Identity maintenance over time based on the on-going re-articulation of core values and commitments within multiple communities of influence.

2) Best understood through an historical analysis of the institutional influences that work inside and outside a particular organization.

3) More dynamic in an organization that maintains competing commitments that pull it in opposing directions. It is in that crucible of angst that debate about the core purposes of the organization breeds. Such angst promises to keep the organization attentive to the influences that shape it. Such mindfulness in turn is crucial in moral reflection and in relating past experiences and values to present pressures (Selznick, 1957; Weick and Sutcliffe, 2001).

4) Sacrificed by both blind utopianism as well as wholesale opportunism. Both the flight to idealism and the adaptation to all environmental changes suggest an organization that is neither mindful of itself nor engaged in the dynamic of re-articulating that self to new realities.

The central contribution of this chapter to this book is to sociologically nuance institutional integrity using LLU as a case. Identity and integrity were shown to be sociological struggles more than simple philosophical debates. Organizational integrity was created, molded, enacted, recounted, and altered by organizational participants fighting for both the survival of an ideology and eventually an organization. It was a struggle to maintain (1) historical roots, (2) financial solvency, and (3) legitimacy in responsiveness to new situations. Scholarship about integrity is therefore less about determining the consistency or rigidity of an organization across time but analyzing that organization's willingness to grow in its application and understanding of its core values in a world of changing expectations. Administrators play a central role in this process as they articulate and apply various institutional sources of legitimacy to justifying organizational practice and mission.

It has been a major assumption of this chapter that advancements in ethical discourse about institutional integrity need to pay more attention to sociology of organizations. Writings on integrity have benefited greatly from philosophical arguments and from fields of psychology. They must now incorporate a layer of complexity that only exists in the literature on the sociology of groups and organizations. Selznick (1992) eloquently argues for the blending of philosophy and social science:

> The distinctive feature of a moral or humanist science is its commitment to normative theory, that is, to theories that evaluate as well as explain....At its best, normative theory is a fruitful union of philosophy and social science. On the one hand, philosophical

acumen is necessary for understanding the complexity and subtlety of basic values and of value related phenomena, such as autonomy, fairness, rationality, love and law. Without sophisticated study of these interdependent variables—including how they have been understood in the history of thought—it is too easy for values to be trivialized or shortchanged. On the other hand, philosophy alone, uninformed by social science, loses touch with empirical contingency and variation and with the insight to be gain from close study of actual experience. (p. xiii)

As the reader will note, I have made a novice's attempt to respond to Selznick's challenge by weaving together historical detail, sociological theory, and philosophical arguments to better show the nature of institutional integrity.

University of Akron
Akron, Ohio, USA

BIBLIOGRAPHY

Brint, S. & Karabel, J. (1991). 'Institutional origins and transformations: the case of American community colleges.' In: DiMaggio, P.J. (ed.), *The New Institutionalism in Organizational Analysis* (pp. 337-360). Chicago: The University of Chicago Press.

Cottrell, R.F., Walters, J.W., Bradley, G. & Rouse, C. (1993, May and June). 'After the divorce: Loma Linda and La Sierra universities.' *Adventist Today, 1,* 13-15.

Covrig, D.M. (1999). *A Case Study of the Organizational History of Loma Linda University: An Examination of Contingency and Institutional Explanations of Development* (Unpublished Doctoral dissertation). Riverside, CA: University of California, Riverside.

Covrig, D.M. (2001a). 'Professional relations: The multiple communities for reform and renewal.' *Professional Ethics, 8*(3/4), 3-38.

Covrig, D.M. (2001b). 'Stability and change in the religious organizational identity of Loma Linda University.' *Research on Christian Higher Education, 8,* 45-68.

DiMaggio, P.J. & Powell, W.W. (1991a). 'Introduction.' In: DiMaggio, P.J. (ed.), *The New Institutionalism in Organizational Analysis* (pp. 1-38). Chicago: The University of Chicago Press.

DiMaggio, P.J. & Powell, W.W. (1991b). 'The iron cage revisited: institutional isomoprhism and collective rationality.' In: Powell, W.W. & DiMaggio, P.J. (eds.). *The New Institutionalism in Organizational Analysis* (pp. 63-82). Chicago: University of Chicago Press.

Dwyer, B. (1992). 'Pursuing that vision thing.' *Spectrum, 22*(3), 3-13.

Flexner, A. (1910). *Medical Education in the United States and Canada; a Report to the Carnegie Foundation for the Advancement of Teaching.* New York: Carnegie Foundation for the Advancement of Teaching.

Gerstner, P. (1996). 'The temple of health: a pictorial history of the battle creek sanitarium.' *Caduceus: A Humanities Journal of Medicine and the Health Sciences, 12*(2), whole issue.

Huntington, S.P. (1997). *The Clash of Civilizations: Remaking of World Order.* New York: A Touchstone Book.

Johns, V. (1984). 'The deans.' In: C. S. Small (ed.), *Diamond Memories: Celebrating the Seventy-fifth Anniversary of the School of Medicine of the College of Medical Evangelists/Loma Linda University* (pp. 63-73). Loma Linda, CA: Alumni Association of the School of Medicine of Loma Linda University.

Knight, G.R. (1995). *The Fat Lady and the Kingdom: Adventist Mission Confronts the Challenges of Institutionalism and Secularization.* Boise, ID: Pacific Press Publishing Association.

Kondrat, M.E. (1999). 'Who is the "self" in self-aware: Professional self-awareness from a critical theory perspective.' *Social Science Review, 73*(4), 451-477.

Ludmerer, K.M. (1985). *Learning to Heal: the Development of American Medical Education.* Baltimore: The Johns Hopkins University Press.

Magan, P.T. (1932 June 9). 'Role of the medical school in the development of character.' *The Medical Evangelist, 18,* 1-4.

May, L. (1996). *The Socially Responsive Self: Social Theory and Professional Ethics*. Chicago: The University of Chicago Press.

Meyer, J.W. & Rowan, B. (1991). 'Institutionalized organizations: formal structure as myth and ceremony.' In: Powell, W.W & DiMaggio, P.J. (eds.), *The New Institutionalism in Organizational Analysis* (pp. 41-62). Chicago: The University of Chicago Press.

Neff, M.L. (1964). *For God and C. M. E.: A Biography of Percy Tilson Magan Upon the Historical Background of the Educational and Medical Work of Seventh-day Adventists*. Mountain View, CA: Pacific Press Publishing Association.

New Loma Linda Corporate Structure Created (1998). [On-line] Available: http://www.llu.edu/directory/lluahsc/news.html

Numbers, R.L. (1973/1992). *Prophetess of Heatlh: Ellen G. White and the Origins of Seventh-day Adventist Heatlh Reform*, revised and enlarged ed., Knoxville, TN: University of Tennessee press.

Numbers, R.L. (1994). 'American medicine comes of age.' In: Judd, W.R. & Butler, J.M. (eds.), *Thirsty Elephant: The Story of Paradise Valley Hospital* (pp. 19-34). National City, CA: Paradise Valley Hospital.

Powell, W.W. & DiMaggio, P.J. (eds.). (1991). *The New Institutionalism in Organizational Analysis*. Chicago: The University of Chicago Press.

Schaefer, R.A. (1995). *Legacy: Daring to Care, the Heritage of Loma Linda Medical Center*. Loma Linda, CA: Legacy Publishing Association.

Schwarz, R.W. (1970). *John Harvey Kellogg, M.D.* Berrien Springs, MI: Andrews University Press.

Schwarz, R.W. (1972). 'The Kellogg schism: the hidden issues.' *Spectrum, 4*(4), 23-39.

Schwarz, R.W. (1979). *Light Bearers to the Remnant: Denominational History Textbook for Seventh-day Adventist College Classes*. Mountain View, CA: Pacific Press Publishing Association.

Schwarz, R.W. (1990a). 'Kellogg vs. the Brethren: his last interview as an Adventist— October 7, 1907.' *Spectrum, 20*(3), 46-62.

Schwarz, R.W. (1990b). 'Kellogg snaps, crackles, and pops; his last interview as an Adventist—Part 2.' *Spectrum, 20*(4), 37-61.

Scott, W.R. (1995a). *Institutions and Organizations*. Thousand Oaks, CA: Sage Publications.

Scott, W.R. (1995b). 'Introduction: institutional theory and organizations.' In: Scott, W.R. & Christensen, S. (eds.), *The Institutional Construction of Organizations: International and Longitudinal Studies* (pp. xi-xxiii). Thousand Oaks, CA: Sage Publications.

Scott, W.R. (1992). *Organizations: Rational, Natural and Open Systems, 3rd ed.* Englewood Cliffs, NJ: Prentice Hall.

Scott, W.R. (1987). 'The adolescence of institutional theory.' *Administrative Science Quarterly, 32*(3), 493-451.

Scott, W.R. & Meyer, J.W. (1991). 'The organization of societal sectors: propositions and early evidence.' In: DiMaggio, P.J. (ed.), *The new institutionalism in organizational analysis* (pp. 108-140). Chicago: The University of Chicago Press.

Scott, W.R., Ruef, M., Mendel, P.J. & Caronna, C.A. (2000). *Institutional Change and Healthcare Organizations: From Professional Dominance to Managed Care.* Chicago: University of Chicago Press.

Selznick, P. (1948). 'Foundations of the theory of organization.' *American Sociological Review, 13*(1), 25-35.

Selznick, P. (1949). *The TVA and the Grass Roots: A Study of Politics and Organization*. Berkeley: University of California Press.

Selznick, P. (1957). *Leadership in Administration*. New York: Harper and Row, Publishers.

Selznick, P. (1992). *The Moral Commonwealth: Social Theory and the Promise of Community*. Berkeley, CA: University of California Press.

Selznick, P. (1996). 'Institutionalism "old" and "new".' *Administrative Science Quarterly, 41*(2), 270-277.

Shryock, E.H. (1984). 'The war years.' In: Small, C.S. (ed.), *Diamond Memories: Celebrating the Seventy-fifth Anniversary of the School of Medicine of the College of Medical Evangelists/Loma Linda University* (pp. 101-109). Loma Linda, CA: Alumni Association of the School of Medicine of Loma Linda University.

Singh, J.V., Tucker, D.J. & Meinhard, A.G. (1991). 'Institutional change and ecological dynamics.' In: DiMaggio, P.J. (ed.), *The New Institutionalism in Organizational Analysis* (pp. 390-424). Chicago: The University of Chicago Press.

Stinchcombe, A.L. (1965). 'Social structure and organization.' In: March, J.G. (ed.), *Handbook of Organizations* (pp. 142-193). Chicago: Rand McNally College Publishing Company.

Turner, J.H. (1997). *The Institutional Order: Economy, Kinship, Religion, Polity, Law, and Education in Evolutionary and Comparative Perspective*. New York: Longman.

Walton, H.M. (1930 December 18). 'Our hopes for Loma Linda.' *The Medical Evangelist, 17,* 1-2.

Weick, K.E. (2001). *Making Sense of the Organization*. Malden, MA: Blackwell Business.

Weick, K.E. & Sutcliffe, K.M. (2001). *Managing the Unexpected: Assuring High Performance in an Age of Complexity*. San Francisco: Jossey-Bass.

White, E.G. (1905). *The Ministry of Healing*. Mountain View, CA: Pacific Press Publishing Association.

White, E.G. (1906 February 1). 'Unscriptural alliances: lessons from the life of Solomon, (Be Ye Separate).' *Review and Herald, 20.*

ANA SMITH ILTIS

ORGANIZATIONAL ETHICS: MORAL OBLIGATION AND INTEGRITY

Organizational ethics in healthcare assesses the obligations of healthcare organizations and addresses how organizations ought to act in particular situations. The past decade has brought increasing attention to organizational ethics from scholars, healthcare executives, the American Medical Association, and the Joint Commission on the Accreditation of Healthcare Organizations.[1] Despite this attention, we lack a shared understanding of what robust moral obligations healthcare organizations bear.[2] There may be a set of minimum obligations borne by healthcare organizations, such as the obligation not to commit fraud. But it may not be possible justifiably to attribute to all healthcare the same obligations to indigent persons, for example. This lack of agreement should come as no surprise given the morally pluralistic composition of our society, a circumstance documented in this volume by Kevin Wm. Wildes, S.J. (2003) and Ronald Arnett and Janie Harden Fritz (2003).

In the absence of a shared understanding of the moral obligations of healthcare organizations, it is helpful to understand them as having two different kinds of moral obligations. First are those justifiably attributed to all healthcare organizations in our society. Second are those that particular healthcare organizations assume. Organizations that are, in the words of Christopher Tollefsen (2003), 'strong institutions' will have a more extensive set of moral obligations than 'weak institutions'. From his perspective, all healthcare organizations share a minimum set of moral obligations, while certain organizations have additional obligations derived from their moral identities. What is morally obligatory for one organization may not be obligatory for another and may be impermissible for yet another organization.

The notion that all healthcare organizations have a minimum set of obligations, while some have additional obligations, raises three questions, all of which are addressed in this volume. First, what are the minimum obligations borne by all healthcare organizations? As one might ascertain from Wildes' as well as Arnett and Fritz's observations regarding moral pluralism, there is likely to be disagreement about the content of these minimum obligations. In this volume, Patricia Werhane (2003), Gerald Logue and Stephen Wear (2003), and Stanley Reiser (2003) offer different accounts of the minimum obligations of healthcare organizations. The

Ana Smith Iltis (ed.), Institutional Integrity in Health Care, 175-182.

differences among the authors suggests that it may be difficult to establish the minimum set of obligations. It is not clear whether any of these three accounts would produce sufficient agreement to allow us justifiably to attribute a common list to all healthcare organizations. The set of shared obligations may be thinner than any of the authors here acknowledges. For example, we may be able justifiably to assert that all healthcare organizations are obligated to respect the forbearance rights of others such that it is impermissible for an organization committed to the beautification of the American population to kidnap and forcibly sterilize those whom its beauty police deem ugly. But we may not be able justifiably to assert that all healthcare organizations must adopt a commitment to providing free healthcare to the poor. The goal of this chapter is not to examine the full range of moral responsibilities attributable to all healthcare organizations. The objective is to demonstrate that there may be minimum obligations justifiably attributable to all healthcare organizations and that particular organizations may be obligated in special ways that extend beyond that set. The precise content of that set of minimum shared obligations is open to debate, as the papers by Werhane, Logue and Wear, and Reiser demonstrate.

Regardless of precisely what obligations are justifiably attributable to all healthcare organizations, it is likely that the list will not include many obligations particular organizations understand themselves as having. For example, while some organizations understand themselves as obligated to support the spiritual well-being of their patients, others may not understand this as an obligation. Organizations who recognize this obligation may rank it differently. A second question raised by this analysis concerns the origin of the additional obligations individual healthcare organizations possess. The papers in this volume by Christopher Tollefsen (2003) and Duane Covrig (2003) are instructive. Some organizations, such as religions institutions, have distinct moral identities such that they possess additional obligations grounded in their moral identities or in what I have called their 'deep moral characters' (Iltis, 2001a and 2001b).[3] In holding itself out as having a particular moral identity, an organization commits to others to fulfill the obligations generated by the identity it affirms. Although the examples in this volume concern religious organizations, there are other moral identities organizations may have that bring them additional obligations.

In attaching this level of authority to an organization's moral identity, it becomes increasingly important that organizations have a strong sense of who they are and that to which they are committed. Although many organizations develop and display mission statements, many of these may be characterized as generic. For example, an organization might say it is committed to "respecting patients' interests and needs" or to "respecting human dignity". These commitments can be interpreted in numerous ways, particularly in a post-modern society such as the one Wildes as well as Arnett and Fritz describe. The contemporary debate on the legalization and moral permissibility of physician-assisted suicide and euthanasia illustrates the radical disagreement in our society regarding the nature of human dignity and what constitutes respect for human dignity. We can expect individuals and organizations to disagree on this and on matters because they have fundamentally different conceptions of morality. Thus it is important for healthcare organizations to develop and articulate more clearly their moral identities if we are to understand these identities as sources of moral obligation. This is not to say all healthcare organizations ought to share a common moral identity. In a morally pluralistic society, organizations should be more clear about their commitments so that we may better understand their obligations and assess their actions.

In the absence of a shared understanding of organizational obligation, a third question emerges: how are we to assess healthcare organizations' actions? If all healthcare organizations had the same moral obligations, then we could compare their actions to that universal set of obligations and assess the extent to which they act as they ought. It is likely that if we had a shared understanding of the full range of healthcare organizations' obligations, we would also have a shared understanding of what is morally good and bad. Thus evaluations of healthcare organizations as having acted in morally good or bad ways most likely would parallel assessments of whether they had fulfilled their obligations. But this is not the case in our society. We have both different understandings of healthcare organizations' obligations and of moral goodness. We are left without a single standard against which we may evaluate the actions of healthcare organizations. We assess the extent to which an organization acts as it ought to by comparing its actions to the full range of obligations it bears, both those it shares with all healthcare organizations and those grounded in its deep moral character. This kind of assessment is separate from an evaluation of an organization as morally good or bad. An organization may hold what some believe are good commitments and others see as bad, yet the organization may fulfill its obligations such that we rightfully assert that it acted as it ought to given its commitments and obligations. Assessments regarding the fulfillment of obligations do not necessarily parallel assessments of organizations as morally good or bad. This is where a particular understanding of integrity becomes important in evaluating the actions of healthcare organizations.

Evaluations of organizational activity turn on the concept of what I have called elsewhere 'moral character integrity' (Iltis 2001a, 2001b) or 'integrity of moral character' (Iltis, forthcoming). To have moral character integrity, an organization must both (1) articulate its moral commitments in ways that are consistent with its deep moral character or moral identity and (2) act in ways that reflect that deep moral character and fulfill the obligations it generates. Organizations that have moral character integrity are not necessarily morally good or morally evil. In fact, some persons may assess the same organization as evil while others hold that the organization is good. Despite such disagreements, judgments of moral character integrity can be consistent. The Roman Catholic healthcare organization that adheres strictly to the *Ethical and Religious Directives for Catholic Health Care Services* (ERDs) (2001) will have moral character integrity. Those who respect the moral commitments reflected in the ERDs are likely to assess that organization as morally good. Others may hold that the organization is, at least to some degree, morally evil.[4]

If moral character integrity does not definitively assess moral goodness, it may seem like a superfluous category of moral evaluation. However, this is not the case, for at least three reasons.

First, in a pluralistic society in which we do not have a shared understanding of morality, moral character integrity allows us to make stronger normative claims regarding an organization's obligations than would otherwise be justifiable. The set of moral obligations we may be able justifiably to attribute to all healthcare organizations is thin. By recognizing organizations as having additional obligations grounded in their moral identities, we expand our understanding of moral obligation. The concept of moral character integrity enables us (1) to attribute to organizations additional moral obligations, namely those that are grounded in an organization's deep moral character, and (2) to assess the extent to which organizations fulfill their obligations. Thus, moral character integrity gives us a richer understanding of moral obligation. There are a number of reasons for which it is desirable that we be able to attribute to individual organizations these additional obligations. Most importantly, when organizations with different moral characters exist, more individuals are able to obtain health care and practice medicine in environments that cohere with their personal values, thus enhancing individual freedom and fulfillment. This is especially important given the circumstance that health care organizations are powerful players in the health care delivery system and there is a lack of agreement on many complex medical-moral problems. This is addressed further in the discussion of the instrumental value of integrity below.

Second, no healthcare organization can commit itself to pursuing all that is good and valuable in this world. Organizations necessarily must choose which goods they will pursue and how they will pursue them. To adopt a particular set of moral aims is to recognize that the limitations all organizations face, such as resource limitations, make it impossible for them to pursue and realize all that is good. The pursuit of a limited range of goods can be important, and moral character integrity permits us to evaluate the fulfillment of these obligations. Once we understand a health care organization as having particular obligations, it is desirable to evaluate the extent to which it satisfies those obligations. Generally we treat moral obligation seriously; we do not recognize agents as bearing moral obligations only to ignore the question of whether they fulfill them. Moral character integrity allows us to judge the extent to which organizations fulfill their particular obligations. Moreover, healthcare organizations often make claims regarding their mission, commitments, values and so on. In doing so, they attribute to themselves moral obligations we might not otherwise have been able justifiably to attribute to them. For example, it may not be possible justifiably to claim that all health care institutions ought to dedicate a certain portion of their activities and income to charity care. However, when an organization takes on this commitment, or has a deep moral character that requires this commitment, we can justifiably assert that it is morally obligated to act in a particular way. This presumes that the organization's deep moral character does not generate obligations that would violate the basic rights we generally ascribe to persons in our society, such as individuals' forbearance rights. Once organizations assume the obligations associated with their particular moral characters, it is desirable that we be able to account for those responsibilities.

Third, moral character integrity is a measure of internal coherence, which we seem both to value for its own sake and as being instrumental to the fulfillment of agents. Our disdain for hypocrisy suggests we generally value moral coherence. We value integrity instrumentally because maintaining integrity is related to being fulfilled. In fact, Baruch Brody holds that integrity is an objective value: integrity is "[a] reflection of the way people *should* relate to their values" (1988, p. 90; italics added). He continues, "We see the formulation of values and goals as a valuable activity but one which would be undercut by a lack of integrity, and we therefore see integrity as something objectively good" (p. 90). Nevertheless, he maintains that integrity is not necessarily a supreme good; integrity is a value among others that can be overridden (p. 54, n. 62). Integrity, he says, "calls upon health care providers and health care recipients to stand firm in their values. It evaluates choices at least in part on the extent to which those choices are consonant with the personal values of both the provider and the recipient of health care" (p. 37).

If we accept that only when persons act on their understandings of the good can they experience fulfillment, and if we recognize that in a morally pluralistic society persons have different conceptions of the good, it is necessary that agents have the moral space necessary, within certain side-constraints, to live what they understand as good and virtuous lives. If agents are unable to pursue their accounts of the good, then they are denied the opportunity to live (what they take to be) full lives. There are individuals who may be able to achieve personal moral character integrity only when there are organizations that reflect their values and within which they can obtain the goods and services they seek. Assessments of moral character integrity enable us to distinguish among various organizations. Such distinctions might be important for patients and health care professionals making decisions about where to seek and provide treatment, respectively. For example, it may be important for some to obtain or provide health care in an environment in which they are not rendered complicit in activities they take to be immoral. Thus, the value of coherent living and integrity for individuals supports the importance of organizational integrity. There are certain goods associated with human flourishing that can be attained only if healthcare organizations have moral character integrity. Organizational moral character integrity makes it possible for individuals to experience fulfillment in the sense of coherence as described by Brody. Thus, all things being equal, it is best for healthcare organizations to maintain their integrity. This does not commit us to accepting that all organizations should always maintain their moral character integrity or that this integrity is always objectively morally good.[5]

The claim that organizational moral character integrity is important should not be equated with the claim that integrity has absolute moral or social value. It will not always be morally good for an institution to maintain its moral character integrity and it will not always be morally bad for an institution to lack full moral character integrity. Nevertheless, it is an important moral concept.[6]

The implications of the moral pluralism described by Wildes as well as Fritz and Arnett in this volume are vast. It is not possible to understand all healthcare organizations as having the same moral commitments. Likewise, it is not possible justifiably to attribute to all persons an identical thick set of moral obligations. We can recognize all healthcare organizations as having a minimum set of moral obligations that is supplemented by the particular obligations derived from organizations' deep moral characters. The shared set of obligations may be thin, perhaps thinner than some of us would prefer. Nonetheless, this circumstance may be inevitable given the degree of moral pluralism we face. In addition to their common obligations, individual healthcare organizations may have additional obligations grounded in their deep moral characters. 'Strong organizations' will have a thicker set of supplemental moral obligations than 'weak organizations'. We can assess the extent to which organizations fulfill their particular moral obligations by evaluating their moral character integrity. Such assessments are not equivalent to assessments of organizations as morally good or morally evil. In the post-modern moment, they are nevertheless important for the reasons outlined in this chapter.

Saint Louis University
Saint Louis, Missouri, USA

NOTES

1. See, for example, Hall (2000), Hofmann and Nelson (2001), Mills, Spencer and Werhane (2001), American Medical Association (2000), Joint Commission on the Accreditation of Healthcare Organizations (1997), Emanuel (1995), Rorty (2000), Schyve (1996), Werhane (2000), and Wildes (1997).
2. Peter French's analysis in this volume (2003) establishes the importance of examining organizational moral responsibility. He explains how and why it is that we may understand organizations, and not only individual persons, as bearers of moral obligations.
3. To avoid radical moral relativism, we must recognize side-constraints that limit the range of deep moral characters it is permissible for healthcare organizations to adopt. To identify and defend a precise set of side-constraints on the deep moral characters healthcare organizations may adopt and hence the commitments they may make would require extensive argument and analysis at this point. The relevant side-constraints would have to include at least those identified by H. T. Engelhardt, Jr.'s principle of permission (1996) which prohibits the use of unconsented to force and requires that agents give their permission before one interferes with them as they peaceably pursue their ends.
4. The organization MergerWatch would likely hold such a view regarding Catholic healthcare organizations for restricting what they take to be basic reproductive rights. MergerWatch is an organization dedicated to "monitor[ing] the threats to reproductive health care from mergers and other health care industry transactions through which restrictive religious rules are imposed on previously secular health care providers and services are banned" (MergerWatch, 2002). They have focused much of their attention on mergers between Catholic and non-Catholic institutions because in such mergers the Catholic facilities generally insist that the new institution refrain from providing services the Catholic organizations find morally objectionable, such as abortions. MergerWatch holds that such healthcare organizations ought not to be permitted to restrict the range of reproductive services they offer and MergerWatch has sought legislative action to prohibit hospitals from limiting the range of reproductive services they offer (Kerry, 1999).
5. The requirement that certain organizations exist in order for particular individuals to live morally coherent and fulfilling lives does not give individuals a positive right to demand that certain institutions exist and be supported so as to ensure their continued existence. It merely gives individuals a forbearance

right, namely the right to be left alone as they associate freely to form and support a particular institution. A detailed exploration of this issue is beyond the scope of this chapter.

6. Not all recognize the concept of organizational moral character integrity as appropriate. Michael Rie, for example, argues that in a secular pluralistic society such as ours, it is permissible for individuals to hold varying moral views that may lead them neither to provide nor to accept particular medical services (e.g., abortion). In choosing to not offer certain services, they would be discriminating against the moral values of others (e.g., those who seek abortion services). Society can accept this type or level of discrimination. However, Rie holds, "society does not accept discrimination on the part of public and private institutions" (Rie, 1991, p. 223). Rie argues that when an organization finds itself holding a position "we" (i.e., Rie and others who share his premises) would deem intolerant and inappropriate in our society, such an institution must withdraw from society. So, the Catholic hospital that refuses to provide abortion services might have to stop providing healthcare altogether, according to Rie. Essentially, Rie recognizes that organizations may have moral commitments, but the moral commitments of individuals carry more moral weight. Healthcare organizations should never maintain their moral character integrity at the expense of individuals. I have already shown that there are important reasons for organizations to maintain their moral character integrity including the possibility that there are cases in which organizational integrity is critical to enabling individuals to maintain their personal integrity and to live lives of moral coherence.

BIBLIOGRAPHY

American Medical Association. (2000). *Working Group on Health Care Organizational Ethics*. Chicago: AMA. Available on-line: http//www.ama-assn.org/ethic/workgroup.htm

Arnett, R.C. & J. M. Harden Fritz (2003). 'Sustaining institutional ethics and integrity: Management in a postmodern moment.' In: Iltis, A. S. (ed.), *Institutional Integrity in Health Care* (pp. 41-71). Dordrecht: Kluwer Academic Publishers

Brody, B. A. (1988). *Life and Death Decision Making*. New York: Oxford University Press.

Covrig, Duane M. (2003). 'Institutional integrity through periods of significant change: Loma Linda University's 100 year struggle with organizational identity.' In: Iltis, A. S. (ed.), *Institutional Integrity in Health Care* (pp. 139-174). Dordrecht: Kluwer Academic Publishers

Emanuel, E. J. (1995). 'Medical ethics in the era of managed care: The need for institutional structures instead of principles for individual cases,' *The Journal of Clinical Ethics*, 6(4), 335-338.

Engelhardt, H. T. Jr. (1996). *The Foundations of Bioethics*. New York: Oxford University Press.

French, P. A. (2003). 'Inference gaps in moral assessment and the moral agency of health care organizations.' In: A. S. Iltis (ed.), *Institutional Integrity in Health Care* (pp. 7-28). Dordrecht: Kluwer Academic Publishers.

Hall, R. T. (2000). *An Introduction to Healthcare Organizational Ethics*. New York: Oxford University Press.

Hofmann, P. B. & W. A. Nelson (eds.) (2001). *Managing Ethically: An Executive's Guide*. Chicago: Health Administration Press.

Iltis, A. S. (2001a). 'Organizational ethics and institutional integrity,' *HEC Forum*, 13(4), 317-328.

Iltis, A. S. (2001b). 'Institutional integrity in Roman Catholic health care institutions,' *Christian Bioethics*, 7(1), 93-104.

Iltis, A. S. (forthcoming). 'Understanding moral obligation in the face of moral pluralism,' *Journal of Value Inquiry*.

Joint Commission on the Accreditation of Healthcare Organizations (1997). *Hospital Accreditation Standards*. Oakbrook Terrace, Illinois: JCAHO.

Kerry, J. M. (1999). 'MergerWatch: It may be coming to a community near you,' *Catholic Health World*, April 1. Available online: www.chausa.org/PUBS/PUBSART.ASP?ISSUE=W990401&ARTICLE=D

Logue, G. & S. Wear (2003). 'The health care institution/patient relationship.' In: A. S. Iltis (ed.), *Institutional Integrity in Health Care* (pp. 99-110). Dordrecht: Kluwer Academic Publishers.

MergerWatch (2002). Home page. Available on-line: www.mergerwatch.org

Mills, A. E., E. S. Spencer, & P. H. Werhane (eds.) (2001). *Developing Organization Ethics in Healthcare: A Case-Based Approach to Policy, Practice, and Compliance*. Hagerstown, Maryland:

University Publishing Group.

National Conference of Catholic Bishops (1995 and 2001) *Ethical and Religious Directives for Catholic Health Care Services* (ERDs). Washington, D.C.: United States Catholic Conference.

Rie, M. (1991). 'Defining the limits of institutional moral agency in health care: A response to Kevin Wildes,' *The Journal of Medicine and Philosophy,* 16, 221-224.

Rorty, M. V. (2000). 'Ethics and economics in healthcare: The role of organization ethics,' *HEC Forum*, 12(1), 57-68.

Schyve, P. M. (1996). 'Patient rights and organization ethics: The Joint Commission perspective,' *Bioethics Forum*, Summer, 13-20.

Tollefsen, C. (2003). 'Institutional integrity.' In: A. S. Iltis (ed.), *Institutional Integrity in Health Care* (pp. 121-137). Dordrecht: Kluwer Academic Publishers.

Werhane, P. (2000). 'Business ethics, stakeholder theory, and the ethics of healthcare organizations,' *Cambridge Quarterly of Healthcare Ethics*, 9(2), 169-181.

Werhane, P. H. (2003). 'Business ethics, organization ethics, and systems ethics for health care.' In: A. S. Iltis (ed.), *Institutional Integrity in Health Care* (pp. 73-98). Dordrecht: Kluwer Academic Publishers.

Wildes, K. Wm. S.J. (2003). 'Institutional integrity in health care: Tony Soprano and family values.' In: A.S. Iltis (ed.), *Institutional Integrity in Health Care* (pp. 29-39). Dordrecht: Kluwer Academic Publishers.

Wildes, K. Wm., S.J. (1997). 'Institutional identity, integrity, and conscience,' *Kennedy Institute of Ethics Journal*, 7(4), 413-419.

NOTES ON CONTRIBUTORS

Ronald C. Arnett, Ph.D., Professor and Chair, Department of Communication and Rhetorical Studies, Duquesne University, Pittsburgh, Pennsylvania, U.S.A.

Duane M. Covrig, Ph.D., Assistant Professor, Department of Educational Foundations and Leadership, University of Akron, Akron, Ohio, U.S.A.

Peter A. French, Ph.D., Lincoln Chair in Ethics and Professor, Department of Philosophy, Arizona State University, Tempe, Arizona, U.S.A.

Janie M. Harden Fritz, Ph.D., Associate Professor, Department of Communication and Rhetorical Studies, Duquesne University, Pittsburgh, Pennsylvania, U.S.A.

Ana Smith Iltis, Ph.D., Assistant Professor, Center for Health Care Ethics, Saint Louis University, Saint, Louis, Missouri, U.S.A.

Gerald Logue, M.D., Professor and Vice-Chair, Department of Medicine, and Co-Director, Center for Clinical Ethics and Humanities in Health Care, State University at Buffalo, Buffalo, New York, U.S.A.

Stanley Joel Reiser, M.D., Ph.D., Griff T. Ross Professor of Humanities and Technology in Health Care, University of Texas Health Science Center at Houston, Houston, Texas, U.S.A.

Christopher Tollefsen, Ph.D., Associate Professor, Department of Philosophy, University of South Carolina, Columbia, South Carolina, U.S.A.

Stephen Wear, Ph.D., Co-Director, Center for Clinical Ethics and Humanities in Health Care, State University at Buffalo, Buffalo, New York, U.S.A.

Patricia H. Werhane, Ph.D., Wicklander Chair of Business Ethics and Director of the Institute for Business and Professional Ethics at DePaul University and Peter and Adeline Ruffin Professor of Business Ethics and Senior Fellow at the Olsson Center for Applied Ethics, Darden School, University of Virginia, Charlottesville, Virginia, U.S.A.

Kevin Wm. Wildes, S.J., Ph.D., Associate Dean, Georgetown College, Georgetown University, Washington, D.C., U.S.A.

INDEX

Philosophy and Medicine

Philosophy and Medicine

21. G.J. Agich and C.E. Begley (eds.): *The Price of Health.* 1986
ISBN 90-277-2285-4
22. E.E. Shelp (ed.): *Sexuality and Medicine.* Vol. I: Conceptual Roots. 1987
ISBN 90-277-2290-0; Pb 90-277-2386-9
23. E.E. Shelp (ed.): *Sexuality and Medicine.* Vol. II: Ethical Viewpoints in Transition.
1987 ISBN 1-55608-013-1; Pb 1-55608-016-6
24. R.C. McMillan, H. Tristram Engelhardt, Jr., and S.F. Spicker (eds.): *Euthanasia and the Newborn.* Conflicts Regarding Saving Lives. 1987
ISBN 90-277-2299-4; Pb 1-55608-039-5
25. S.F. Spicker, S.R. Ingman and I.R. Lawson (eds.): *Ethical Dimensions of Geriatric Care.* Value Conflicts for the 21th Century. 1987 ISBN 1-55608-027-1
26. L. Nordenfelt: *On the Nature of Health.* An Action-Theoretic Approach. 2nd, rev. ed. 1995 SBN 0-7923-3369-1; Pb 0-7923-3470-1
27. S.F. Spicker, W.B. Bondeson and H. Tristram Engelhardt, Jr. (eds.): *The Contraceptive Ethos.* Reproductive Rights and Responsibilities. 1987
ISBN 1-55608-035-2
28. S.F. Spicker, I. Alon, A. de Vries and H. Tristram Engelhardt, Jr. (eds.): *The Use of Human Beings in Research.* With Special Reference to Clinical Trials. 1988
ISBN 1-55608-043-3
29. N.M.P. King, L.R. Churchill and A.W. Cross (eds.): *The Physician as Captain of the Ship.* A Critical Reappraisal. 1988 ISBN 1-55608-044-1
30. H.-M. Sass and R.U. Massey (eds.): *Health Care Systems.* Moral Conflicts in European and American Public Policy. 1988 ISBN 1-55608-045-X
31. R.M. Zaner (ed.): *Death: Beyond Whole-Brain Criteria.* 1988
ISBN 1-55608-053-0
32. B.A. Brody (ed.): *Moral Theory and Moral Judgments in Medical Ethics.* 1988
ISBN 1-55608-060-3
33. L.M. Kopelman and J.C. Moskop (eds.): *Children and Health Care.* Moral and Social Issues. 1989 ISBN 1-55608-078-6
34. E.D. Pellegrino, J.P. Langan and J. Collins Harvey (eds.): *Catholic Perspectives on Medical Morals.* Foundational Issues. 1989 ISBN 1-55608-083-2
35. B.A. Brody (ed.): *Suicide and Euthanasia.* Historical and Contemporary Themes.
1989 ISBN 0-7923-0106-4
36. H.A.M.J. ten Have, G.K. Kimsma and S.F. Spicker (eds.): *The Growth of Medical Knowledge.* 1990 ISBN 0-7923-0736-4
37. I. Löwy (ed.): *The Polish School of Philosophy of Medicine.* From Tytus Chałubiński (1820–1889) to Ludwik Fleck (1896–1961). 1990
ISBN 0-7923-0958-8
38. T.J. Bole III and W.B. Bondeson: *Rights to Health Care.* 1991
ISBN 0-7923-1137-X

Philosophy and Medicine

39. M.A.G. Cutter and E.E. Shelp (eds.): *Competency*. A Study of Informal Competency Determinations in Primary Care. 1991   ISBN 0-7923-1304-6
40. J.L. Peset and D. Gracia (eds.): *The Ethics of Diagnosis*. 1992
          ISBN 0-7923-1544-8
41. K.W. Wildes, S.J., F. Abel, S.J. and J.C. Harvey (eds.): *Birth, Suffering, and Death*. Catholic Perspectives at the Edges of Life. 1992 [CSiB-1]
          ISBN 0-7923-1547-2; Pb 0-7923-2545-1
42. S.K. Toombs: *The Meaning of Illness*. A Phenomenological Account of the Different Perspectives of Physician and Patient. 1992
          ISBN 0-7923-1570-7; Pb 0-7923-2443-9
43. D. Leder (ed.): *The Body in Medical Thought and Practice*. 1992
          ISBN 0-7923-1657-6
44. C. Delkeskamp-Hayes and M.A.G. Cutter (eds.): *Science, Technology, and the Art of Medicine*. European-American Dialogues. 1993   ISBN 0-7923-1869-2
45. R. Baker, D. Porter and R. Porter (eds.): *The Codification of Medical Morality*. Historical and Philosophical Studies of the Formalization of Western Medical Morality in the 18th and 19th Centuries, Volume One: Medical Ethics and Etiquette in the 18th Century. 1993   ISBN 0-7923-1921-4
46. K. Bayertz (ed.): *The Concept of Moral Consensus*. The Case of Technological Interventions in Human Reproduction. 1994   ISBN 0-7923-2615-6
47. L. Nordenfelt (ed.): *Concepts and Measurement of Quality of Life in Health Care*. 1994 [ESiP-1]   ISBN 0-7923-2824-8
48. R. Baker and M.A. Strosberg (eds.) with the assistance of J. Bynum: *Legislating Medical Ethics*. A Study of the New York State Do-Not-Resuscitate Law. 1995
          ISBN 0-7923-2995-3
49. R. Baker (ed.): *The Codification of Medical Morality*. Historical and Philosophical Studies of the Formalization of Western Morality in the 18th and 19th Centuries, Volume Two: Anglo-American Medical Ethics and Medical Jurisprudence in the 19th Century. 1995   ISBN 0-7923-3528-7; Pb 0-7923-3529-5
50. R.A. Carson and C.R. Burns (eds.): *Philosophy of Medicine and Bioethics*. A Twenty-Year Retrospective and Critical Appraisal. 1997 ISBN 0-7923-3545-7
51. K.W. Wildes, S.J. (ed.): *Critical Choices and Critical Care*. Catholic Perspectives on Allocating Resources in Intensive Care Medicine. 1995 [CSiB-2]
          ISBN 0-7923-3382-9
52. K. Bayertz (ed.): *Sanctity of Life and Human Dignity*. 1996
          ISBN 0-7923-3739-5
53. Kevin Wm. Wildes, S.J. (ed.): *Infertility: A Crossroad of Faith, Medicine, and Technology*. 1996   ISBN 0-7923-4061-2
54. Kazumasa Hoshino (ed.): *Japanese and Western Bioethics*. Studies in Moral Diversity. 1996   ISBN 0-7923-4112-0

Philosophy and Medicine

Philosophy and Medicine

KLUWER ACADEMIC PUBLISHERS – DORDRECHT / BOSTON / LONDON